Military Transformation and Strategy

This book explores the idea of a "revolution in military affairs" (RMA), which underpins the transformational agenda of the US military, and examines its implications for smaller states.

The strategic studies literature on the RMA tends to be American-centric and directed towards the strategic problems of the US military. This volume seeks to fill the gap in the literature and establish an intellectual framework that can assist other, smaller powers in their respective approaches to this issue.

The book does so in three main sections: Part I focuses on questions of transformations in strategy and war; Part II explores transformations in operations; while Part III examines possible impediments to an RMA.

This book will be of much interest to students of Military Studies, Asian Studies, Strategic Studies and International Relations in general.

Bernard Loo is Assistant Professor at S. Rajaratnam School of International Studies, Singapore.

Contemporary security studies

Series Editors: James Gow and Rachel Kerr
King's College London

This series focuses on new research across the spectrum of international peace and security, in an era where each year throws up multiple examples of conflicts that present new security challenges in the world around them.

NATO's Secret Armies
Operation Gladio and terrorism in
Western Europe
Daniele Ganser

The US, NATO and Military Burden-Sharing
*Peter Kent Forster and
Stephen J. Cimbala*

Russian Governance in the Twenty-First Century
Geo-strategy, geopolitics and
new governance
Irina Isakova

The Foreign Office and Finland 1938–1940
Diplomatic sideshow
Craig Gerrard

Rethinking the Nature of War
*Edited by Isabelle Duyvesteyn and
Jan Angstrom*

Perception and Reality in the Modern Yugoslav Conflict
Myth, falsehood and deceit
1991–1995
Brendan O'Shea

The Political Economy of Peacebuilding in Post-Dayton Bosnia
Tim Donais

The Rift Between America and Old Europe
The distracted eagle
Peter H. Merkl

The Iraq War
European perspectives on politics,
strategy, and operations
*Edited by Jan Hallenberg and
Håkan Karlsson*

Weapons Proliferation and War in the Greater Middle East
Strategic contest
Richard L. Russell

Military Transformation and Strategy

Revolutions in military affairs and small states

Edited by Bernard Loo

LONDON AND NEW YORK

First published 2009
by Routledge
2 Park Square, Milton Park, Abingdon, Oxon OX14 4RN

Simultaneously published in the USA and Canada
by Routledge
270 Madison Ave, New York, NY 10016

Routledge is an imprint of the Taylor & Francis Group, an informa business

Typeset in Times by Wearset Ltd, Boldon, Tyne and Wear
Printed and bound in Great Britain by TJI Digital, Padstow, Cornwall

British Library Cataloguing in Publication Data
A catalogue record for this book is available from the British Library

Library of Congress Cataloging in Publication Data
p. cm. – (Contemporary security studies)
Includes bibliographical references.
1. National security–Developing countries. 2. Military art and
science–Technological innovations–United States. 3. Special operations
(Military science)–United States. 4. Military doctrine–United States.
5. Strategy. 6. States, Small. 7. World politics–21st century. I. Loo,
Bernard Fook Weng.
UA10.5M533 2008
355'.03301724–dc22   2008008936

ISBN10: 0-415-42110-1 (hbk)
ISBN10: 0-203-89182-1 (ebk)

ISBN13: 978-0-415-42110-2 (hbk)
ISBN13: 978-0-203-89182-7 (ebk)

Contents

Contributors

David J. Betz is Senior Lecturer in the Department of War Studies, King's College London.

Malcolm Brailey is an analyst at the Australian Department of Defence. He was formerly an Associate Research Fellow for the Institute of Defence and Strategic Studies, Rajaratnam School of International Studies, Nanyang Technological University, Singapore, when he contributed this chapter.

Eliot Cohen is the Robert E. Osgood Professor of Strategic Studies, the Director of the Strategic Studies Program and the Director of the Philip Merrill Center for Strategic Studies at the Paul H. Nitze School of Advanced International Studies (SAIS) at Johns Hopkins University.

Christopher Coker is Professor of International Relations, LSE and Visiting Professor at the Staff College, Oslo.

James R. FitzSimonds is Professor in the Warfare Analysis and Research Department at the Naval War College. He is a graduate of the U.S. Naval Academy and holds an SM degree from MIT.

Joshua Ho is Senior Fellow at the S. Rajaratnam School of International Studies, Nanyang Technological University.

Elizabeth Kier is Associate Professor at the University of Washington.

Bernard Loo is Assistant Professor at S. Rajaratnam School of International Studies, Nanyang Technological University.

Thomas G. Mahnken currently serves as the US Deputy Assistant Secretary of Defense for Policy Planning. He previously served as a Professor of Strategy at the US Naval War College and a Visiting Fellow at the Philip Merrill Center for Strategic Studies at the Johns Hopkins University's Paul H. Nitze School of Advanced International Studies (SAIS).

Ron Matthews is Professor of Defence Economics and Academic Leader of the Masters in Defence Administration Course in the Department of Defence Management and Security Analysis, Cranfield University, UK Defence Academy.

Paul T. Mitchell is Associate Professor at the Department of Defence Studies, Canadian Forces College, Toronto

Manjeet Singh Pardesi is a doctoral candidate at the Department of Political Science at Indiana University. He was an Associate Research Fellow for the Institute of Defence and Strategic Studies, Rajaratnam School of International Studies, Nanyang Technological University, Singapore, when he contributed his chapter.

Nancy Roberts is Professor in the Graduate School of Business and Public Policy (Professor of Strategic Management) and in the School of International Graduate Studies (Professor of National Security Affairs) at the Naval Postgraduate School, Monterey.

Introduction

Revolutions in military affairs: theory and applicability to small armed forces

Bernard Loo

Is there a revolution in military affairs (RMA) underway? What does this putative RMA actually mean? What are its key features and characteristics? Finally, and this is the question that this volume seeks to address, what does this RMA mean for small states? These are questions that, largely, are not very well answered in the existing RMA literature. This is particularly true of the last question – it is a reflection of the wider strategic studies literature that tends, in large part, to address the strategic conditions and problems of great powers, and which often does not translate very well in its application to smaller armed forces with much more limited capabilities than their great power counterparts.

The immediate problem, especially for the last question – and this is the research dilemma that drives this volume – is that there is little information about how small states can engage with this alleged RMA. It starts from the observation that the wider strategic studies literature tends to be great power-centric. Furthermore, this alleged RMA is driven almost exclusively by developments in the United States (US) military, whose defence expenditure is of a scale and size that far outstrips any other state, even its closest competitors. Finally, this RMA is still a fairly recent phenomenon, with a body of literature that remains relatively young and less than fully developed. This volume seeks to begin to address this lacuna. This obviously is a step into the unknown. For small states wanting to engage in this RMA, it therefore seems unavoidable that the journey begins with an examination of how the great powers engage in this RMA.

Understanding RMA theory

A revolution in military affairs is ultimately the idea that dramatic changes in any number of variables of war – such as weapons technologies, to cite but one – lead to fundamentally and radically different approaches to the entire military structure and its modus operandi. There are any number of definitions of the RMA, but this volume suggests that these various definitions can be encapsulated in the above statement. In some instances, such as the RMA manifested in the Napoleonic wars, it could be changes in the makeup of society (Knox 2001: 57–73). It could be due to changes in military doctrines and tactics, such as the

RMA of fourteenth-century England (Rogers 2001: 15–34). Alternatively, it could be due to changes in the structure of the military industrial complex, such as the American Civil War (Grimsley 2001: 74–91). In all these previous RMAs, the predominant drivers of revolution were different variables of war – one was doctrinal, one was societal, yet another was military–industrial.

One key question is whether or not an RMA is currently underway. There is as yet no consensus view of this issue. This is a question that is ultimately harder to resolve. How one identifies a revolution currently underway is ultimately dependent on the markers of radical change that one seeks. What analysts tend to do is to focus on certain aspects of the military – whether in doctrines, organizational structures, operations and tactics – and to look for indicators of change. How radical these changes are then perceived is a matter of personal choice thereafter. Searching for signs of an RMA is a highly unscientific business.

Characteristics of the current RMA

This existing concern about the existence of an extant RMA notwithstanding, there is a body of opinion that insists that an RMA is currently extant (Adams 1998; Arquilla and Ronfeldt 1997; Cohen 1999; Cohen 1996: 37–54; Freedman 1998; Laird and Mey 1999; O'Hanlon 2000; Owens 2000; Schneider and Grinter 1995; Thomas 1997; Toffler and Toffler 1993). From this body of opinion, a predominant understanding of this RMA is beginning to emerge – that the current RMA is essentially technical in nature, and that it first manifested itself during the 1991 Gulf War against Iraq (Kendall 1992: 23–30; Mazaar 1993: 17–39). As for the alleged current RMA, the key drivers of this revolution would appear to be centred on technological changes in the computing and communications spheres.

The driving force behind this RMA is information processing, which manifests itself in three key aspects: information dominance, precision weaponry, and joint-service operations. Information dominance promises to disperse the fog of war for friendly commanders and thicken it for the enemy. A network of sensor systems will gather data in order to generate real-time, continuous, target-quality information on all significant enemy assets. Advanced command, control and communications (C^3) will then transform this data, along with information on friendly forces, into a single real-time near-perfect picture of the battlespace available to all commanders. This will enable commanders to target, shape and distort the enemy's understanding of the same battlespace (Hewish 1994: 51).

Second, improvements in precision targeting render increasingly likely the prospects of single-shot kills. Together with information dominance, precision weapons promise near-certain destruction, which for friendly forces, promises to make warfare much more efficient than ever, and to make desired strategic outcomes almost guaranteed. The RMA-ed military can therefore utterly dominate the battlespace. It is the ability to guarantee the destruction of the most important targets (that is, intensive destruction) that determines military success, rather than indiscriminate saturation bombings (that is, extensive destruction) as was witnessed in the Second World War. This might allow a relatively small force to defeat a much larger enemy formation. Long-range standoff-range weapon systems may be particularly important because they can strike from locations that are beyond the range of most hostile weapons (Henry and Peartree 1998).

Finally, information technology allows for the networking of all aspects of the military organization – that is, jointness – that generates a net effect far greater than the sum of its parts. Networking facilitates a common organization-wide awareness of the battlespace, and this promises to generate force multiplier effects as a result of allowing any emerging target to be engaged and destroyed by precision weapons launched from any platform that is within range. If all three aspects of the RMA hold true, this will likely result in an armed forces that is increasingly lethal and capable of levels of precise destruction previously unattainable (Chandler 1998).

The inevitability of military change

Yet, military change is almost certainly inevitable, as societal, political, and military–technological conditions change. This RMA is driven by the cutting edge of technological change, which means that at some point, countries have to decide whether or not they want to engage in their own RMA agendas. This is the argument that technological dynamics impose by themselves a sort of irrefutable logic. It is not unlike the decision to buy a new car – after a point, maintaining an old car may simply no longer make economic sense – older, less fuel-efficient engines in an era of ever-rising fuel costs makes running an old car increasingly expensive. Furthermore, the lack of spare parts makes maintaining the engine of the car increasingly difficult. The fundamental issue here is that no military force can afford to be static in nature and capability. Military and military-related technologies necessarily change, and existing capabilities and equipment eventually become obsolete. To remain relevant and effective, all military forces have to undergo periodic change, both in terms of their hardware and capabilities as well as in terms of their doctrines and strategies.

Barry Buzan (1987; also Buzan and Herring 1998) hints at this ever-changing nature of the military, in his notion of the arms dynamic. The focus of Buzan's analysis, however, is unhelpfully limited – it focuses on the changing nature of armaments, with insufficient focus on the changing nature of doctrines and tactics, force structures and military organizations. RMA theory at least attempts to take the analysis further, and examine changes in these other aspects of the military. Even so, applying this argument to small states is inherently problematic. The technological dynamic is changing at an accelerating pace. Taking into consideration the ever-increasing costs of new military technologies, the limited resources of small states may not be sufficient for these actors to keep pace with the military technology dynamic. This then presents serious challenges for small states wanting to keep pace with new military technologies.

Engaging in the RMA

Strategy and the great power–small state dichotomy

The question is, of what relevance is this alleged RMA for non-superpowers, small states in particular. The current RMA, if it exists at all, was born out of a unique set of geopolitical circumstances, and designed to meet a unique set of strategic conditions. It was the attempt by the US to answer the challenges posed by what was, in the 1970s and 1980s, a numerically superior Warsaw Pact conventional capability in the central European theatre of operations. The technologies that underpin this current RMA sought to provide a technological solution to the problem of an imagined conventional force-on-force encounter between the US and its North Atlantic Treaty Organization (NATO) allies on the one hand, and the former Soviet Union and the Warsaw Pact on the other (Murray and Knox 2001: 1–4). What relevance would such an RMA have for other countries, first not endowed with the material wherewithal of the US and second not likely to face similar force-on-force encounters with their putative adversaries? Even if the current RMA is relevant to non-superpowers, is this relevance across-the-board? Or, more likely, are certain aspects of the current RMA directly relevant? These are difficult questions that defy attempts to provide definitive answers.

What this volume presupposes is that strategy for great powers is fundamentally different than that of small states. It is not to say that the concepts of strategy – understood here as at least the threat, if not the actual use, of force being applied to achieve the interests of the state – differ from great to small powers (Gray 1982: 7, 1988: 30; Klein 1994: 2). However, the difference is not in strategy and its fundamentals, but in the way policy-makers, scholars and strategists approach the application and implementation of strategy.

Small countries with small populations face the imperative of constantly reviewing their defence force structures, the military capabilities that their military establishments deem to be necessary, their doctrines and operational procedures. Nevertheless, it is worth remembering that it is still a matter of debate as to what this process will mean for the very nature of war and strategy, let alone military operations and tactics. Military organizations may find it extremely difficult to not only absorb new capabilities, but adapt to new modus operandi as well. This problem is only exacerbated by the extremely high costs of modern weapons systems, which being already a problem for the United States can only be an even greater problem for smaller states. Finally, there is the issue of relevance to prevailing strategic conditions. Given current strategic conditions, the RMA may create an armed force that will be excellent for strategic tasks that are rapidly becoming less and less likely.

Affording the RMA – transformation as journey or as destination?

A common theme in US approaches to military transformation is the idea that transformation is a journey, not a destination. Cebrowski writes, "First and foremost, transformation is a continuing process. It does not have an end point. Transformation anticipates and creates the future and deals with the co-evolution of concepts, processes, organizations, and technology" (Cebrowski 2003: 8). There is of course the difficulty of determining what is then the "right path" of this journey of military transformation. As Cebrowski further notes,

> [N]ew capabilities rarely outperform the old when they first appear. Often this is because the technologies involved may be only nascent, and the required doctrinal or organizational changes are incomplete. So there is a danger in culling new capabilities too soon ... Eventually, such efforts will render previous ways of warfighting obsolete and change the measures of success in military operations.

Implicit here is the notion that experimentation and failure are necessary elements of this process of transformation. This process of experimentation and failure is after all an inevitable aspect of scientific progress. For any state wanting to engage with this process of military transformation, it seems almost inescapable that they will have to endure this process of experimentation and failure before finding their own "right path" towards military transformation. However, and this is a theme that Ron Matthews explores in this volume, the current transformation agenda of the US military is an expensive business, and the US military spends at a level that is beyond the financial resources of all other states. Put simply, small and medium powers wanting to engage in military transformation may not be able to afford to "get it wrong." That being the case, small and medium powers may have to first decide what this process of military transformation will bring them to, its strategic value given the kinds of security challenges they each face, the economic resources that they can devote to transformation, and the kind of armed forces they will have post-transformation.

In other words, for small and medium powers, military transformation has to be as much a journey as a destination. This automatically renders the US experience of military transformation problematic as a guide for small and medium powers. The US currently offers the only template of military transformation that is available for all other states wanting to engage in military transformation; and yet, the strategic circumstances facing the US are fundamentally unique, utterly different from those that other small and medium powers face. Not only is the strategic environment for each power fundamentally different from the strategic environment of the US, the US simply has resources that are beyond those of any other extant state. It is imperative for small and medium powers wanting to engage in transformation to study the US experience. However, this has to be done with care. Engaging with the RMA and the transformation agenda, therefore, necessarily requires that small and medium powers critically examine the US experience for the correct lessons for their respective transformation agendas.

As suggested above, affording the RMA is going to be a difficult challenge for small and medium powers. This is an issue that Ron Matthews examines in

this volume. Value for money – translated into economy, efficiency and effectiveness – is becoming an increasing priority for all government agencies, especially the armed forces, which is almost always the single largest consumer of
public resources. For small states like Singapore, this is an even more difficult
problem to overcome. For the modern military organization, this aim translates
into the minimization of the so-called "tail end" of the military (the non-combat,
support side) to the "teeth end" (the combat side). Matthews argues that value
for money has become even more important in the current RMA, given the
increasing high costs of emerging weapons systems and technologies. Current
combat systems are simply a lot more expensive than their predecessors, and
this has resulted in a kind of structural disarmament, whereby states and military
organizations can afford ever-decreasing numbers of new weapons systems and
platforms. Nowhere is this more prevalent than in air power, as successive generations of combat aircraft are becoming more and more expensive (Kirkpatrick
and Pugh 1983). Singapore recently announced its decision to buy 12 F-15T
combat aircraft (locally designated F-15SG) as a replacement for its existing
fleet of 40 A-4SU. The idea is that the 12 F-15SGs will have sufficient combat
power that more than compensates for the smaller number of combat platforms.
Nevertheless, it makes the loss of even a single platform all the more costly, and
potentially strategically disastrous.

Overcoming organizational obstacles

For most states embarking on their RMA agendas, these technologies possess a
significant potential to negatively impact on organizational structures and work
processes. For mature conventional military organizations, the change might be
even more important, dramatic and disturbing – leading to quite plausibly fundamental changes in the structure of the military organization and the doctrines
with which these military organizations engage in armed conflict.

   Even the most modern of military organizations is essentially an industrial-
era organization, characterized by centralized controls and processes manned by
a large body of staff, fairly rigid hierarchies, and high degrees of functional spe-
cializations. Such organizations regard information as a means to an end,
whereas the current information revolution sees information as an end in itself.
Martin Van Creveld (1985: 247–248) argues that such industrial-era organizations tend to suffer from "information pathology" – one example being how
rapidly growing message traffic in Vietnam clogged the extant military signals
networks, with little or no ability to differentiate between low and high-priority
signals. Such an industrial-era model may have been good where the competitive advantage lay in maximizing output (and information was the tool for maximizing output), but may be ill adapted to situations where the focus of attention
shifts to information as the end product (Scott 1987: 76–92). For the military,
the problem may be analogous to other organizations in supplying an
information-driven marketplace. Focused on optimizing the placement of firepower on "targets" – that is, on the production process – an industrial-model

organization could easily fail to address the combat equivalent of the customer – *the enemy*. The industrial approach to war regarded the enemy merely as a factor in the process, never as a human being with an independent will. For all the destruction it wrought, firepower applied along industrial lines seldom destroyed – or even attacked – the most important target: the enemy's will to resist.

Implications for military operations

Facing new security challenges

However, conventional military operations may very well prove to be the least likely operational scenario that modern militaries find themselves having to face.

Operations other than war

However, conventional wars are being increasingly supplanted by insurgencies and civil wars (Kaldor 1999). The soldier today is more likely to be deployed in such non-traditional (and unfamiliar) roles including peace operations, humanitarian and disaster relief, and counter-terrorism – so-called operations other than war (OOTW) – rather than the traditional defence of the state against foreign invasion. One priority item has been peace operations, ranging from peace enforcement to peacekeeping. Certainly, military organizations have been increasingly involved in these types of operations (Gordenker and Weiss 1991: 2). As of October 2006, the United Nations has deployed 80,976 military and police personnel in 18 different peace operations around the world.

It is not only United Nations-sanctioned peace operations that have increased in numbers. What has also increased is the number of so-called low-intensity conflicts; indeed, since the Second World War, only 12 per cent of conflicts can be classified as high-intensity conflicts (Reid 1998: 28). One study actually goes so far as to argue that the majority of conflicts in the medium term will be low-intensity intra-state in nature, rather than high-intensity and inter-state (Metz and Millen 2003: 13). Yet another study shows that the probability of states becoming involved in inter-state conflicts has dropped from 1 in 28 from 1918–1941, to 1 in 167 from 1945–1990, and 1 in 250 from 1991–2003 (Holsti 2006: 136). As the US experience in Vietnam and the Soviet experience in Afghanistan show, a military organization skilled at high-intensity conventional operations can flounder in the less familiar terrain of low-intensity conflict.

However, these OOTW (operations other than war) roles are inherently problematic (Doyle 1998: 8–12), and require skill sets different from those demanded by conventional military operations, which typically do not occupy very much attention in the training regimes of modern militaries (Moskos *et al.* 2000). Indeed these OOTW skills might even run against the grain of the more traditional warrior skills a soldier is supposed to have. In peace operations, for instance, the mission aim is to avoid conflict and casualties; soldiers are expected to display non-threatening behaviour, which runs against the grain of their training. There is increasing recognition of the different challenges that peace operations impose on militaries, and the acceptance of the need for specialized training – such as the creation of the Pearson Peacekeeping Centre in Canada.

Fighting the Global War on Terror

It is always tempting to employ the military in counter-terrorism efforts. Military organizations almost always have the necessary manpower and the skills for counter-terrorism. Inasmuch as terrorist bases and facilities can be located, through the use of intelligence, surveillance and reconnaisance (ISR) capabilities that are the purview of military organizations, these can be attacked and destroyed by either the careful insertion of trained military operatives or the precise application of standoff-range firepower. Even the more passive counter-terror measures – such as the guarding of critical infrastructure – resonates with that most mind-numbingly boring yet necessary of military tasks – the provision of guard and sentry posts. Tapping this reservoir of manpower resources for counter-terrorism efforts therefore appears to make sense. Furthermore, it appears at first glance that the RMA does hold promise for counter-insurgency and counter-terrorism operations, that with modern command, control and communications systems, information can be made more available to commanders to provide them with a clearer situational picture of the area of operation, facilitating a faster operational cycle (Berkowitz 2003).

That counter-terror missions are going to be more frequent in the near term seems an almost universally held opinion now. But these will likely be terrorists quite unlike their predecessors – rather, these will be combatants who are as computer-savvy as their counterparts. The information technology revolution has become a double-edged sword. Terrorists – or if nothing else, super-empowered and disenfranchised angry young men armed with credit cards, computer notebooks and modems – are also able to access technology that is commercially available to carry out operations against states that have strong military forces. This is the argument that Thomas X. Hammes (2004) makes in his book, *The Sling and the Stone*. United States military forces in Iraq are currently facing such insurgents. This presents a dilemma to all militaries as postwar operations are certainly more complicated than ever before.

However, a cautionary note is needed. The principles of counter-terrorism are not entirely consonant with the principles of conventional warfare. The military mindset focuses on proactive problem-solving – find the problem and then fix or destroy it. It is reflected in two axioms – "Never send in a man when a bullet will do"; and "Firepower is cheaper than manpower." Success can then be easily determined – at least, if the threat emanates from another state's regular conventional forces. However, firepower is a whole lot more expensive in the highly politicized milieu of counter-terrorism, where the critical effort resides in so-called "hearts and minds" measures.

Even when military force can be brought to bear in counter-terrorism, for example, when terrorist bases are located, the application of firepower has to be very carefully calibrated, so as to not incur unnecessary levels of destruction, especially collateral damage. In any case, terrorists and their bases are not so easy to locate and destroy, otherwise the problem would not be as intractable as it seems. Counter-terrorism more typically involves passive security measures – the guarding of critical infrastructure and installations, which more resemble law enforcement and policing. In both law enforcement and counter-terrorism, the measure of success is reflected in the absence of incidents. Restraint in the use of force is desirable in counter-terrorism, but this may run against the grain of the military mindset.

This volume seeks to provide a picture of how radical changes might be underway in selective aspects of military organizations and their respective modus operandi. It attempts to do so in three sections, each focusing on specific aspects of the military enterprise. Part I focuses on questions of transformations in strategy and war. In Chapter 1, Eliot Cohen argues the case that a revolution in military affairs is currently underway, at least in the armed forces of the United States of America. At one level, differences in the appearance of soldiers is already manifest – soldiers from the Second World War through to the Vietnam War looked more or less alike in terms of their uniforms and the equipment they carried. The modern American soldier, however, now wears different uniforms that come with bullet-proof vests, have radios built into their helmets, carry a range of devices that allow them to operate at night, and see over the hill with small unmanned aerial vehicles. He further argues that the lethality of war has increased, and concurrently the economy of war. Recent conflicts have shown that, at least at the tactical levels, battles can be waged faster and more precisely. Christopher Coker examines the bio-technological transformations that are currently underway, primarily in the United States armed forces, and the ethical challenges that will have to be negotiated through as bio-technological advances threaten to undermine the very human aspect that wars have always encompassed.

Part II turns the attention towards transformations in operations. Joshua Ho directs his attention to the theory of effects based operations (EBO). He decries notions that EBO represents a fundamental break from military planning in the past, but argues instead that ultimately, all military planning is about the generation of effects beyond the immediate confines of the battlefield. Manjeet Pardesi looks at the increasing emphasis on unmanned aerial vehicle (UAV) technologies, and argues that it is probably erroneous to think of UAVs as being able to completely supplant manned aerial platforms. UAVs, he argues, do possess certain advantages over manned platforms, but suffer from the lack of tactical flexibility and full situational awareness. Malcolm Brailey examines the impact that RMAs have had on special operations forces (SOF), and argues that there is clear evidence that the current RMA has had a significant impact on SOF operations, doctrines and tactics. David Betz examines the impact of RMA technologies on military operations other than war (MOOTW). He argues that although certain RMA capabilities can be usefully applied to MOOTW, MOOTW thoroughly encapsulates the Clausewitzian war–politics nexus, and concludes that information dominance cannot be seen as a substitute for sound political judgement.

Part III turns the attention turns towards the possible impediments to an RMA. Thomas Mahnken and James Fitzsimmons examine US military officer attitudes towards the issue of military transformation. They argue that the apparent concrete evidence of RMA superiority from past conflicts such as the 1990 Gulf War notwithstanding, there remain pockets of scepticism. Such attitudinal obstacles will have to be overcome if the RMA project is to reach its fruition.

References

Adams, James (1998) *The Next World War: Computers are the Weapon and the Front Line is Everywhere*, New York: Simon & Schuster.

Arquilla, John and Ronfeldt, David (eds) (1997) *In Athena's Camp: Preparing for Conflict in the Information Age*, Santa Monica: RAND.

Berkowitz, Bruce (2003) *The New Face of War: How War will be FOUGHT in the 21st Century*, New York: The Free Press.

Cebrowski, Arthur K. (2003) *Military Transformation: A Strategic Approach*, Washington, DC: Office of Force Transformation, The Pentagon.

Chandler, Robert W. (1998) *The New Face of War: Weapons of Mass Destruction and the Revitalization of America's Transoceanic Military Strategy*, McLean, VA: Amcoda Press.

Cohen, Eliot A. (1996) "A Revolution in Warfare," *Foreign Affairs*, 75(2): 37–54.

Cohen, William S. (1999) *Annual Report to the President and the Congress*, Washington, DC: The Pentagon.

Doyle, Michael W. (1998) "Discovering the Limits and Potential of Peacekeeping," in Olara A. Otunnu and Michael W. Doyle (eds) *Peacemaking and Peacekeeping for the New Century*, Oxford: Rowman & Littlefield.

Freedman, Lawrence (1998) "The Revolution in Military Affairs," *Adelphi Paper*, 318.

Gordenker, Leon and Weiss, Thoman G. (1991) *Soldiers, Peacekeepers and Disaster*, London: Macmillan.

Gray, Colin S. (1982) *Strategic Studies: A Critical Assessment*, London: Aldwych Press.

Gray, Colin S. (1988) *The Geopolitics of Superpower*, Lexington: University Press of Kentucky.

Grimsley, Mark (2001) "Surviving Military Revolution: The U.S. Civil War," in MacGregor Knox and Williamsom Murray, (eds) *The Dynamics of Military Revolution 1300–2050*, Cambridge: Cambridge University Press.

Hammes, Thomas X. (2004) *The Sling and the Stone: On War in the 21st Century*, St. Paul: Zenith Press.

Henry, Ryan and Peartree, C. Edward (eds) (1998) *The Information Revolution and International Security*, Washington, DC: Center for Strategic and International Studies.

Hewish, Mark (1994) "Fishing in the Data Stream," *International Defense Review*, July.

Holsti, Kalevi J. (2006) *The Decline of Interstate War, or The Waning of Major War*, London: Routledge.

Kaldor, Mary (1999) *New and Old Wars: Organized Violence in a Global Era*, Cambridge: Polity Press.

Kendall, Frank (1992) "Exploiting the Military Technical Revolution: a Concept for Joint Warfare," *Strategic Review*, Spring: 23–30.

Kirkpatrick, David and Pugh, P. (1983) "Towards the Starship Enterprise – are the Current Trends in Defense Unit Costs Inexorable?" *Aerospace* (May 1983): 16–23.

Klein, Bradley (1994) *Strategic Studies and World Order: The Global Politics of Deterrence*, Cambridge, New York and Melbourne: Cambridge University Press.

Knox, MacGregor (2001) "Mass Politics and Nationalism as Military Revolution: The French Revolution and after," in MacGregor Knox and Williamsom Murray (eds) *The Dynamics of Military Revolution 1300–2050*, Cambridge: Cambridge University Press.

Laird, Robin F. and Mey, Holger H. (1999) *The Revolution in Military Affairs: Allied Perspectives*, Washington, DC: National Defense University Institute for National Strategic Studies.

Mazaar, Michael J., Shaffer, J. and Ederington, B. (1993) *Military Technical Revolution: A Structural Framework*, Final Report of the Study Group on the Military Technical Revolution, Washington, DC: Center for Strategic and International Studies.

Metz, Steven, and Millen, Raymond (2003) *Future War/Future Battlespace: The Strategic Role of American Landpower*, Carlise, PA: Strategic Studies Institute, US Army War College.

Moskos, Charles, Williams, John Allen and Segal, David R. (eds) *The Postmodern Military: Armed Forces after the Cold War*, Oxford: Oxford University Press, 2000.

Murray, Williamson and Knox, MacGregor (2001) "Thinking about Revolutions in Warfare," in MacGregor Knox and Williamsom Murray (eds) *The Dynamics of Military Revolution 1300–2050*, Cambridge: Cambridge University Press.

O'Hanlon, Michael (2000) *Technological Change and the Future of Warfare*, Washington, DC: Brookings Institution Press.

Owens, William (2000) *Lifting the Fog of War*, New York: Farrar, Straus & Giroux.

Reid, Brian (1998) *The Nature of Future Conflicts: Implications for Force Development*, Camberley, UK: Strategic and Combat Studies Institute.

Rogers, Clifford J. (2001) "'As if a New Sun had Arisen:' England's Fourteenth-century RMA," in MacGregor Knox and Williamson Murray (eds) *The Dynamics of Military Revolution 1300–2050*, Cambridge: Cambridge University Press.

Schneider, Barry R. and Grinter, Lawrence E. (eds) (1995) *Battlefield of the Future: 21st Century Warfare Issues*, Air War College Studies in National Security 3, Maxwell Air Force Base, Alabama: Air University.

Scott, W. Richard (1987) *Organizations: Rational, Natural, and Open Systems* 2nd edn, Englewood Cliffs, NJ: Prentice-Hall.

Thomas, Keith (ed.) (1997) *The Revolution in Military Affairs: Warfare in the Information Age*, Canberra: Australian Defense Studies Centre.

Toffler, Alvin and Toffler, Heidi (1993) *War and Antiwar: Survival at the Dawn of the 21st Century*, Boston: Little, Brown & Company.

Van Creveld, Martin (1985) *Command in War*, Cambridge, MA: Harvard University Press.

Part I
Transformations and strategy

1 Change and transformation in military affairs

Eliot Cohen

For nearly 20 years military analysts have talked, alternately, of a "military technical revolution," a "revolution in military affairs," and most recently "military transformation." But has a radical shift in military affairs occurred? Or have we merely seen normal, evolutionary processes of change at work in the conduct of war? After all, over a quarter of a century – a period that spans entire military careers – has passed since the first discussions of Soviet-style military technical revolution appeared. It is reasonable to ask: did it happen? Or has the art of war, as many historians and soldiers long suspected, merely witnessed the usual processes of steady change and evolution?

The notion of a revolution in military affairs (RMA) has had numerous proponents, and many variants, but a common ground with four key aspects can be identified. One, the advent of superior information technology and weapons of precision has vastly enhanced the power of advanced military forces. Two, it is possible to conduct operations that do not follow classic patterns of advancing along fronts with discernible lines and rear areas. Three, the new technologies make numbers and platforms far less important than networks and communications. Finally, military operations now aim at defined effects rather than attrition of enemy forces or occupation of ground.

Revolutions in military affairs – four critiques

This summary, of course, simplifies what has been, in fact, a cluster of theories with many variants. Still, four general problems with this cluster of theories stand out: (1) the abstraction of RMA theorizing from the world of geopolitics; (2) its focus on technology at the expense of the softer aspects of military affairs (organization, doctrine, manpower, etc.); (3) a tendency to depict transformation as something that happens top-down, rather than bottom-up; (4) a failure to look at the response to RMA-type capabilities on the part of weaker opponents. Ironically, it is only by looking at these four aspects of military affairs in our time that we can truly estimate the possibilities of military transformation.

Abstraction from geopolitics

Most of the discussion of military transformation has said little about geopolitics. Thus, William Owens' notion of a "system of systems" that could integrate communications, intelligence, and precision strike systems, enabling the United States (US) military to observe, detect, target and destroy any object within a 200-mile by 200-mile box was, at one level, a politically neutral technical discussion of capabilities (Owens and Offley 2000). But of course it only made sense as something that a global, indeed, a hyperpower, might have the interest and resources to do.

Other depictions of revolutionary change rested equally on unstated geopolitical assumptions. Thus, Soviet writings from the late 1970s through the mid-1980s about a military technical revolution were couched in terms of "reconnaissance strike complexes" – but the underlying issue was the American development of the capability to annihilate mass armoured formations moving towards the inter-German border. That capability, which the Soviets were not at all certain they could counter, put at risk Soviet operational concepts for war with the West, and beyond it, the entire structure of Soviet strategy.

The tools of war – technology, organization, operational concepts – have embedded within them assumptions about what war is, how it can and should be waged, by whom and against whom it will be conducted. Much of the thinking about transformation of modern militaries has a distinctly US feel to it – because it is produced by Americans, who understandably think about the world in an American way. The geopolitical assumptions behind this thinking are that the US is a country with global interests, shifting threats, and vast resources. Chinese, Australian, Singaporean, or Israeli circumstances are quite different, even if they proceed from a similar technological base, and hence one should expect, and indeed hope for, concepts of transformation that vary by country, and may take very different forms.

The classic cases of the development of armoured doctrine and organization in the interwar period, and of advanced concepts for carrier aviation during the same time illustrate this point. Germany developed the Panzer division because of its location in Europe, its military traditions, and its aggressive ideology. Japan and the US pushed carrier aviation because they anticipated war with one another in the Pacific. The Japanese did not develop panzer divisions, and the Germans did not develop aircraft carriers, even though the technologies were well within the reach of both countries.

To set itself on a sounder footing, then, theorizing about an RMA has to begin by acknowledging that its roots lay in a world divided between two superpowers; where potential for global conflict was quite small, but the calculations involved in preparing for it were politically important; where the most important theatre of war was central Europe. The kind of war envisioned by the original Soviet and American theorists was of a type familiar from the first half of the twentieth century, and in some respects even the nineteenth century: a period of rising tensions, swift mobilization, and decisive combat. That world has vanished, being replaced by an unprecedented international system dominated by one hyperpower, with some rising competitors, whose ideas of war are altogether different, where war may or may not take the form of swift decision, but where in any case follow through is likely to be characterized by succeeding waves of violence and struggle in a variety of spheres, including cyberspace and the contest for public opinion.

Today, the US has the only full spectrum military in the world. It alone can afford and deploy the full panoply of military strength; it also is the only military that must respond to a global range of threats and potential. For evidence of the resources at its disposal, consider the 2003 supplemental bill to pay for American expenses in Afghanistan and Iraq: some US$87 billion on top of what had already been spent to wage war in both places. That increment was about eight times the size of the Australian defence budget, roughly twice the British defence budget. It is sometimes suggested that the US looms so large because its allies have been unwilling to commit forces by its side. But in fact, when considered in proportion, population, size of military and defence budget, countries like Poland and Italy have made commitments that would be comparable to the US committing 40,000 or 50,000 troops to a given operation. It is the *absolute* size of the US that is so staggering.

Of course, soldiers feel pinched for funds, and complain about it. But the fact is that the US can pour vast sums into defence, and does. The American research and development budget is larger than the British defence budget, to take just one example. And that means staggering investments in a whole range of areas (which will not, in all likelihood, be sustainable even for the US), for example, large investments in two new and very different fighters, while modernizing the existing tactical fighter force, open up an entirely different kind of airborne platform in the shape of unmanned aerial vehicles (UAV).

The American RMA is designed to fit US strategic needs – the need for power projection, quick wins, low casualties, and the flexibility to move from one theatre to another. Other countries have very different purposes. China, for example, seeks to use advanced technologies for the purpose of coercing, and possibly occupying Taiwan and projecting power into the South China Sea. With a panoply of unmanned systems – ballistic missiles, long-range, high-speed cruise missiles, advanced torpedoes, mines – together with a small number of advanced aircraft and ships, China wishes to develop the ability to isolate, intimidate, or even occupy Taiwan even in the face of US opposition. Its military has the advantage of a focused set of strategic and operational challenges, and even if it is by all measures inferior to those of the US. Thus, in some ways the People's Liberation Army (PLA) may achieve its RMA in ways that trump those of the US, while looking quite different from them.

In a similar vein, the Israel Defense Forces (IDF) has probably led the world, including the US, in the development of military technology and operational concepts for urban guerrilla warfare. Operation Defensive Shield in March and April of 2002, during which Israeli forces swept into the West Bank to root out terrorist organizations, was, from the military–technical point of view, a remarkable success. In the face of a hostile and armed population the Israelis were able to achieve their objectives without suffering excessive losses, or inflicting large numbers of civilian casualties. Inter-squad radios, aerostats, and sophisticated surveillance technology combined with old-fashioned bulldozers allowed the Israelis to do something that was hitherto thought well-nigh impossible: to conduct successful urban operations under conditions of limited force.

The theory of the RMA cannot be simply a US theory, and it should not be an apolitical strategic concept, because politics will always come back in through the back door, as it were. The challenge is twofold: for the tailoring of the theory of transformational change to local circumstances, while at the same time describing the broader trends as comprehensively as possible.

Technology over organization

The technologies that drive military transformation are now embedded in our daily lives. We think nothing of telephone services that recognize human voices, computers that allow us to examine, select, and pay for hotel rooms half a world away, cheap handheld devices that not only tell us where we are on the surface of the earth to a resolution of ten metres, but tell us how to get from where we are across town – if necessary by talking to us. The information technologies above all underpin the transformation of war, and it may be precisely because they are so familiar that we fail to see how much change they have brought about. It was, after all, only a few years ago that economists were insisting that the application of information technology in the business world had yielded no increases in productivity – arguments that were, of course, written on personal computers and subsequently posted on the Internet.

Still, by itself technology does not yield transformational change, particularly in the military world. In business it is different: competitive pressures mean that companies that do not adopt electronic scanning quickly run up large inventory costs. Military organizations can do business the old way longer, particularly if they are not operational. It is no coincidence that the most active militaries in the world – the American, British, and Israeli among them – have also been some of the leaders in transformation. Even then, however, activity can reinforce some kinds of conservatism precisely because a mistake in innovation can bring with it human and political costs of a kind rare or unknown in the commercial world.

A lag between the appearance of useful technologies and their mature application in organizations and concepts of operations is not surprising. What *is* surprising in the theory of the RMA is the relative paucity of attention to the human software, as it were, as opposed to the material hardware. Arguably, however, the concepts and organizations are now catching up with technology. It is only in the last few years, for example, that the US Army has begun the routine use of tracking technology that allows commanders to know where all their vehicles are at any time and to share data by intranet and digital maps. The vast increase in precision air power – foreshadowed more than a generation ago in the large-scale use of guided weapons in Vietnam – has finally made a large difference on the battlefield with the use of reliable and secure communication from ground observers, laser designators, and precision navigation systems. As a result of all these changes, the US Army has begun to study the notion of taking the standard mechanized infantry division of three brigades and turning it into five brigades. More important yet is the evolution of the brigade – a combined arms organization, much like an armoured cavalry regiment – into the central unit in the US Army. The Afghan War of 2001 and the Iraq War of 2003 highlighted these changes, which, though far from complete were well under way. Other aspects of the American military are changing too: the introduction of submarines that can deliver large numbers of long-range cruise missiles, for example. In short, although the reaction of many military observers to American success in Iraq in 1991 was one of surprise and, indeed, admiration (genuine, fearful, or grudging, as the case may be), the real growth in US military power may only now be coming into existence, as a new generation of officers takes charge of the armed forces. This post-Vietnam generation is far more confident and aggressive in its use of information technologies and far more "joint" in its outlook than its predecessor.

For other countries change will be more painful. In the case of European militaries, for example, the challenges begin at a mundane level, simply in the creation of the smaller, professional armed forces that seem to be mandated by the new technologies. Conscription is withering away slowly in Europe, and the result in countries like Germany is not merely a bloated and ineffective military, but an organization that cannot fully exploit the potential of the new technologies. Even those countries with professional armed forces, however, often find themselves lacking some of the basic prerequisites for effective military power – adequate stocks of precision weapons or night vision devices, or the logistical infrastructure to sustain operations at some considerable distance from the homeland. For a time, then, the gap between the US and other militaries is likely to grow, not diminish.

Transformation as top down or bottom up

One characteristic of much of the writing about military transformation has been its implicit assumption that change would come from above, that is, from enlightened senior leadership imposing different ways of war on recalcitrant bureaucracies. When the military transformation debate became briefly a matter of political interest in the 2000 US presidential campaign, both candidates spoke as if change would have to come from the top. And, indeed, Donald Rumsfeld, who may well be the most transformation-oriented Secretary of Defense in US history, began his tenure with a series of centralized studies that smacked very much of change imposed from above. Even as some of that tendency to centralism has abated, the US military created an office of transformation in the Office of the Secretary of Defense, and other countries have been tempted to take a similar course.

There is no doubt that some changes do indeed have to come from the top. But RMA theorists may have overestimated the degree to which enlightened senior leadership could, by itself, remake the armed forces, as opposed to creating conditions that would by themselves foster change. Throughout most of military history, to include the current period, change tends to come more from below, from the spontaneous interactions between military people, technology, and particular tactical circumstances. The critical question is whether an organization is capable of taking those changes and adopting them widely. Thus, for example, the development of infiltration tactics during the First World War was most successful in Germany, where a strong general staff was able to learn from front line innovators and disseminate new ideas rapidly (Lupfer 1981). Similarly, it was not the rather conservative leadership of the German army in the interwar period that gave birth to the Panzer division – but it *was* that leadership that supported innovators with whom it in fact disagreed.

Today, the greatest changes in war are brought about by spontaneous innovation in reaction to tactical problems, as when, during the 2003 Iraq War, a Marine company commander was able to view on a laptop computer images from a fighter-bomber flying overhead, looking down the street in an Iraqi town. It is the urge to tinker and experiment – by having a UAV pass targeting information to an F-15E Strike Eagle, for example, or by giving personal digital assistants to everyone on board a warship – that creates new ways of war, not comprehensive plans created from the top and passed down to a reluctant organization. At the heart of real change in military affairs is the notion of a "learning organization," which is something quite different from a brilliant organization. This in turns requires an organizational culture that encourages experimentation and does not punish the failures that innovation invariably brings about. These qualities, in turn, rest on fundamental attributes of societies that reflect themselves in their militaries. For that reason, societies that do not see occasional failure as calamitous, that are willing to allow juniors to overcome or contradict seniors, and that do not value "face" or reputation excessively are likely to succeed in transforming themselves. It is no coincidence that the US, a country that views bankruptcy as a learning experience, has created one of the most innovative militaries in the world. This of course points to the challenge of military transformation for societies that might prefer to bring change about by enlightened diktat.

Responses by weaker opponents

The enemy never really figured very much in the RMA debate, and this may have been the worst mistake of all. While American theorists and foreign imitators spoke in abstract terms of 200-mile by 200-mile boxes, sensor-to-shooter links, and dominant battlespace awareness, the reality was, as we have seen, one of the growths of US military power. In the real world of geopolitics, however, countries and non-state actors respond to US dominance, including its superiority in high-technology warfare, in a variety of ways.

For countries like China the response has been one of selective transformation with a view to neutralizing those American advantages that most endanger them. This has taken a variety of forms, to include work on computer attacks, weapons that disable electronic systems, or that can negate key American advantages such as stealth aircraft. The idea is not to counter American strength across the board, but rather to negate American capabilities that could threaten Chinese superiority where it counts most: in the Taiwan Straits and adjoining waters. For the moment, only China, of the militarily advanced powers seems to have gone down this route; Russia might desire to do so, but remains too impoverished to do more than market the products of its once flourishing technological research programmes. This approach of selective transformation for local purposes seeks to capitalize on asymmetries inherent in the US global posture. The US must, to achieve its political objectives, project power abroad; it must manoeuvre substantial forces. The Chinese approach has been to emphasize destructive power over manoeuvre, and to acquire only very limited power projection capabilities.

A second kind of response is the quest for weapons of mass destruction (WMD). Nuclear and biological weapons (chemical weapons only secondarily) seem to some states as a great equalizer: realizing that they cannot hope to thwart US power directly, they seek, by acquisition of exceptionally powerful weapons to deter American attack and exercise influence, by blackmail or implicit threats, over their neighbours. The programmes of Iraq, North Korea, and Iran must be understood in this light. This response, however, has its own limitations. In the case of Iraq the appearance and perhaps the reality of such programmes increased the willingness of the US to attack. In the other cases, active WMD programmes have not improved relations with the US, and have as yet to translate into a real deterrent against American action. Furthermore, it is clear that some of the American reaction to such programmes is, in fact, the use of advanced military technologies to counter such weapons by active defence or possibly, in the future, by pre-emption. The voluntary surrender of such programmes by Libya suggests that the combination of economic pressure and military threat by the US makes the utility of such deterrent strategies questionable.

The third, and perhaps the most worrisome, counter to US military superiority is the use of asymmetric techniques of terror – techniques that could become even more lethal if, as appears likely, they become married with WMD. The new breed of Islamic terrorists are technologically sophisticated, using the Internet for clandestine communications, for example, while relying on the willingness of some operatives to commit suicide for a cause. Whether in spectacular strikes against homeland targets (as on 11 September 2001 in New York, or 11 March 2004 in Madrid), or in attacks on military targets such as the USS Cole attack of 12 October 2000, or simply in hit and run attacks on American forces in Iraq, the modern terrorist has managed to combine ingenious uses of modern technology with old-fashioned fanaticism.

Again, this counter to the advanced military potential of the US is far from decisive. The qualities of modern societies that allow them to succeed in high technology warfare give them many tools to respond to asymmetric threats. These include a wide range of homeland security technologies and measures, from explosives detection sensors to biometric identification and to surveillance and reconnaissance capabilities that are entirely applicable to an insurgency. Again, the ability of an advanced military to adapt to new threats is in some ways more impressive than the skills it brings to bear to begin with. The conversion of American heavy divisions (the 4th Infantry or 1st Armoured divisions, for example) to counterinsurgency forces in Iraq is a good example of this. The technology–organization complex that makes the American military good at intense warfare can and does contribute to some (not all) aspects of counterinsurgency – realistic training, for example, or database management, or the ability to operate stealthily at night.

Has there been a revolution in military affairs?

Has a revolution occurred? Some military commentators have been doubtful (for instance, Biddle 2002). More importantly, how would one judge a transformation of military affairs if one had actually occurred? Three tests seem reasonable: do military forces look fundamentally different from what they were in the past? Are the processes of battle different than they once were? And are outcomes also different?

Transformed militaries

One way of addressing the first question is by looking at soldiers themselves. A soldier serving in the Vietnam War did not look fundamentally different from his father who served in the Second World War. The uniforms and protective gear were the same: green cotton fatigues and steel helmet; the weapons similar, an M-16 being neither more accurate nor more lethal than an M-1, although delivering a higher rate of fire; the various accoutrements had seen modest improvements (from aluminium to plastic canteens, better but still bulky hand-held radios, C-rations in cans and intermittent mail service from home). Today's soldier looks altogether different. His protection now includes a helmet that can turn bullets (unlike the steel helmet) and a vest that will do the same. His weapons have sights that make a mediocre shot a marksman at most engagement ranges. He can see at night with the aid of devices that were just beginning to become available in the 1970s, and indeed, prefers to fight at night when possible. His radio is now inside his helmet, and rather than being confined solely to officers, has been distributed to every soldier in a squad. Even the amenities of daily life are different, from the rations he eats to the hydration pack he wears in lieu of carrying a canteen.

The modern high tech soldier is a far more lethal creature than his predecessor. Not only are his weapons much better; with radio, laser designator, and precision navigation he can bring down vast amounts of firepower from aircraft, artillery, and missiles. His officers can see over the proverbial hill with small UAVs transmitting pictures in real time, and coordinate fire accordingly. Above him are aircraft that routinely deliver precision firepower of a kind qualitatively different from anything known in the history of warfare to this point. Even at the level of the common infantryman, then, armed forces have changed. To be sure, many of his circumstances have remained the same – physical exertion and hazard. But in almost every other respect he is no longer what he was, a mass consumable of war, but rather a highly trained specialist.

War remains a rough and dirty business. But the forces that wage it are fundamentally unlike their predecessors. Technology, including, in particular, information processing technology is now pervasive in every aspect of war. One result is that only in very special circumstances (Israel being the most notable case) can conscript forces prove competitive in large-scale warfare. For most countries they are simply too expensive to be very good, with the result that developed countries have either shifted to volunteer militaries, or simply gotten out of the business of having sophisticated armies. The highly paid professional militaries of today increasingly live like what they are; members of the middle class, with perquisites and benefits comparable to, and in many ways superior to those of members of a large corporation. And indeed, corporations have come to take on some of the roles once filled only by soldiers. Logistical bases are constructed and administered by contractors; military and paramilitary organizations trained by them; sophisticated weapons not only developed and built, but also maintained by them. Even the application of force has become, in some modest measure, the business of business, as the private security business has boomed.

Transforming battle outcomes

If armed forces are different, so too are battle outcomes. The discrepancy between advanced military powers and others has yielded victories that were easily anticipated, but were far more lopsided than had been expected (Posen 2003). Beginning with the overwhelming Israeli victory over Syrian air defences in the Bekaa Valley in 1982, through the Iraq War of 1991, the Bosnia and Kosovo Wars of 1995 and 1999 through the Afghanistan War of 2001 and the Iraq War of 2003, Western powers, and the US above all has won smashing victories at costs in human life that are remarkably low.

There have been various attempts to explain these outcomes – the ineptitude of particular military organizations, a systematic underestimation of air power, and sheer luck.[1] The sequence of such lopsided successes however, over a period of 20 years, and indeed the growth in the disproportionality of such outcomes requires explanation. Indeed, when one considers that beyond real battles one must consider the battles that never happened, i.e., those that were deterred by the superiority of advanced military organizations (another Arab–Israeli war during the second *Intifadah* being the most plausible), the phenomenon looks all the more striking.

Much of this has to do with the military unipolarity of the world. Without a Soviet Union to provide abundant first-rate military hardware and advice to opponents of the US and its allies, the military resources of the US must inevitably predominate. But that alone is not the whole story. The Soviet

Union's collapse came from the absurdity of its economic system and the demoralization of its peoples and its rulers. It coincided with, and may have been related to, the gradual draining of its military self-confidence. In the mid-1970s the Soviet Union could pride itself on catching up to the US in many spheres of military activity, to include the development of advanced technology. By the late 1980s the gap had begun to open again, and even if the Communist Party had not imploded then, the Soviet military would have found itself slipping further and further behind. Weapons systems requiring sophisticated and autonomous operators, exacting maintenance, and industrial technology from advanced materials to clean rooms, were beyond the reach of a society built on command, subordination, and often abusive treatment of subordinates.

It is a cliché but true nonetheless that armed forces reflect their societies. The phenomenon sometimes called "the digital divide" separates advanced societies (or communities within those larger societies) that have successfully embraced the information revolution and those that have not, and perhaps may not do so any time soon. There is today a conventional military competence divide that is no less great, and that will persist at least as long as those other divides do.

Transforming the nature of battle

Finally, does battle itself look different? This is more difficult to judge than the other two criteria for transformational change. Combat has always been an extremely variegated experience. Anyone who has had the experience of interviewing veterans knows that one sometimes has the sense of speaking to soldiers who have fought not only in different battles, but in altogether different wars. Still, four generalizations may be made.

First, the core experiences of fear and adrenaline-induced excitement have not changed: they are probably recognizably those described in *The Iliad*, and will continue to be true so long as men wage war. The phenomenon of post-traumatic stress, first recognized during the First World War and since recognized as a psychological fact of war continues. From here on, however, change becomes increasingly obvious.

Second, modern warfare is increasingly warfare without fronts. The combination of long-range air power, equally long-range missile and rocket fires, and special operations forces has gradually eroded the idea of front lines and more or less secure rear areas, except, perhaps, on a continental scale. Whether the enemy be guerrillas or conventional forces, the instinct will be to bypass the increasingly thin conventional forces to strike at vulnerable supply lines or command posts. Thus, in the 2003 Iraq War, the Iraqis seem to have decided from the outset to wage a war in depth, and, indeed, did so with some success.

Third, the central activity of combat is beginning to change. Reading accounts of the fighting in Iraq – be it in the intensive combat operations of March 2003, or the subsequent contest of ambushes and urban firefights since – much of the combat looks as it ever did. It is close, brutal, and chaotic. Thus it ever was and in many respects thus it ever will be, at least insofar as the average infantryman is concerned. But subtle changes have already occurred, particularly insofar as advanced militaries are concerned, that make even close combat different.

For one thing, the well-equipped and highly trained soldier described above has an advantage quite different from that obtained in the past, in close combat against his irregular or merely mediocre opponent. Superior morale, discipline, and training made a big difference in the past, and sometimes decided the issue. Today, technology plays an important role. The soldiers of sophisticated armies are much harder to kill, have far better sensors, and far more lethal resources at their disposal than their opponents. They can call in accurate firepower in the third dimension, using precision fires from aircraft (manned and increasingly unmanned), as well as indirect weapon systems such as guns and mortars. Their sensors give them an edge in night combat, and even during the day, small unmanned vehicles, superior mapping and navigation technology make them far more formidable, even in the urban maze.

In suitable terrain the change is all the more striking. The rise of modern special operations forces stems not merely from their skills in combat, but from their ability to wield massive fires that they can accurately and reliably direct on to targets. This change was most noticeable in the 2001 Afghan War, but continued in Iraq as well. The result is to force opponents into dispersed guerrilla and even terrorist tactics. The appearance of routine precision fire on the battlefield, at levels from the long-range sniper rifle accurate at distances of 500 metres and beyond, to guided cannon and mortar rounds, bombs and missiles, does not remove the importance of manoeuvre, but in some ways diminishes its salience. In the shifting balance between fire, manoeuvre, and shock, the three constants of land warfare, fire is of increasing importance. Similar developments have been noted for some time in air and naval warfare as well.

Finally, at the higher levels of combat, a major change is apparent in the automation of many functions of command, and in a vast increase in the amount of reliable information at command posts in the field, and the networking of that information within units, in a theatre, and even globally. While many of the functions of command remain the same, to include the unchanging requirements to select, motivate, evaluate, replace and promote subordinates, the modern commander operates in a different communication environment than ever before.

Conclusion

There is, then, reason to think that a major change – call it transformation or not – in warfare has occurred. It stems from fundamental changes in international politics (the rise of the US, the temporary absence of competing powers, and the collapse of order in various parts of the world), and the maturation of the information technologies. The interaction between politics, society, and technology will continue, and new forms of warfare will evolve, but it is at least conceivable that the new forms of warfare will persist for some decades, at least until biological technologies compel yet another shift in the means of combat. The dominance of the US and its preferred modes of high technology warfare do not necessarily mean a peaceful world, or success for Washington – in some respects just the reverse, as opponents seek irregular or unconventional responses to American strength. But the RMA is here to stay.

Note

1 John Keegan, "So the bomber got through to Milosevic after all," *Daily Telegraph*, 6 April 1999.

References

Biddle, Stephen (2002) "The New Way of War?" *Foreign Affairs*, 81(3): 138–144.

Lupfer, Timothy (1981) *The Dynamics of Doctrine: The Changes in German Tactical Doctrine During the First World War*, Fort Leavenworth, KS: US Army Command and General Staff College.

Owens, William A. and Offley, Edward (2000) *Lifting the Fog of War*, New York: Farrar, Straus & Giroux.

Posen, Barry (2003) "Command of the Commons: The Military Foundation of U.S. Hegemony," *International Security*, 28(1): 5–46.

2 Biotechnology, military transformation and the future of war

Christopher Coker

It is often said that science fiction is a genre that combines the cognitive, (the rational, scientific) and estrangement (translated as alienation from the familiar and the everyday). In reality, most science fiction writing is an extension, or extrapolation of the present. If it were only concerned with estrangement we would not understand it. And if it were only about cognition it would be a work of science, not science fiction. It is the combination of both that allows science fiction to challenge the ordinary – or what we take for granted.

Now, if we had been looking at the future of war in 1908 we would have learned much from a book written by H.G. Wells in which he predicted the coming of an atomic war. If we want to glimpse the future of war today where do we go?

We could start with Orson Scott Card's *Ender's Game* (1991) in which we find the training of soldiers in their early years takes the form of "games" in the Game Room. The government has taken to breeding military geniuses and then training them in the art of war. Another influential sci-fi book is Leo Frankowski's *A Boy and His Tank* (1999) which tells of a group of colonists on a planet combining Virtual Reality with tank warfare, a world in which warriors bond with their tanks, and their tanks with them. One of the most telling lines in the book is: "Kid, if your tank is loyal you don't have to be!" Card's novel is now taught on the leadership course of the United States (US) Marine Corps University at Quantico. Frankowski's book was proofread by a solider from Company C – Task Force, 1–32 Armour First Cavalry Division whilst he was deployed in the desert, waiting for "Desert Storm."

The two novels illustrate the two revolutions that are transforming the face of war. Of the three revolutions that have shaped war since 1945, the atomic led war into an endgame, nuclear deterrence – though this it must be said may be more apparent than real. We have entered what some commentators call "the second nuclear age," an age in which nuclear weapons may be used for the first time. But for the advocates of the RMA, and most other writers on future warfare, it is the information and biotechnology revolutions that hold out the greatest promise of playing the game by other rules. For western societies that are forever sensible of public distaste for war both revolutions may offer the chance to play the game a little longer.

Although the focus here is biotechnology it is becoming clear that the two revolutions are not distinct. Digital biology is likely to be the key to the future in every walk of life. Indeed, the decoding of the human genome would have been impossible without the information technology (IT) revolution. For genetic manipulation requires the decoding and recombining of the information codes of living matter, which is made possible only by an exponential increase in processing power. Conversely, the language of the IT age has been significantly influenced by nature. What the genome project reveals is that we have almost the same number of genes as the chimpanzee. What differentiates us from other animals is the networking and recombining capacity of our brain cells, through millions of electrochemical connections. It would seem that in terms of networking and feedback loops (the basis of cybernetics) the human brain is similar to the Internet in our computer-run societies.

In time scientists who understand the processes of nature (especially those who know how complex adaptive systems work) will be able to build computers that can *evolve* (rather than solve) most conceivable problems. In computer programming "evolutionary algorithms" (programs that permit things to evolve in computer space) are dictating the pace of change. In an attempt to create more complex computer brains scientists are also studying complex neural networks in the human brain in the expectation of constructing "digital chromosomes" with many of the same features as our own DNA.

When it comes to war, digital biology is already redrawing the rules. Instrumentally, war is being defined in biological terms. Existentially the warrior too may soon be enhanced through cyborg technologies and genetic re-engineering which promise him – or her – the chance to breed out the imperfections of the past; the chance *to breed true*.

It is a challenge which is typical of the age in which we live, one in which the biological is privileged more and more over the cultural. It would seem, as evolutionary psychologists tell us, that human behaviour is far more genetically determined than we had previously thought, and that by enhancing, modifying, or altering our genes we may well be able to enhance the things we do well, and have always done well as a species. One of the things we have done particularly well over the centuries has been war, and there is nothing to suggest we will be going out of the war business, indeed quite the reverse.

The challenge of the future

Biological frontiers

Biology (but not biotechnology) has already changed the way in which we look at military operations, as well as the use of force in post military contexts. For some years now, "The Marine Corps After Next" (MCAN) Branch of the Marine Corps Warfighting Laboratory has been exploring what it calls a "biological systems inspiration" for future warfighting. MCAN argues that in approaching war in Newtonian terms (i.e., ordered and predictable), we have been ignoring the more likely understanding of war as a complex system; i.e., an "open ended, parallel and very sensitive to initial conditions and continued 'inputs.' Those inputs are the 'fortunes of war'" (Metz 2000: 34). If this is correct, then the key characteristic we need for military success is smallness, dispersion, autonomy and adaptability.

The characteristics of an adaptable, complex system closely parallel biology. To deal with the biological is to do least damage to the "environment," understood broadly as the social, political as well as ecological context within which war is fought. When the term "ecology" was first coined in the 1860s it described the holistic study of living systems interacting with their environment. Ecologists look at communities of organisms, patterns of life, natural cycles and demographic changes. And this is precisely what a new generation of the military is doing as well.

When we turn to biotechnology, however, its influence is likely to be found not in the instrumental so much as the existential dimension of war. This too is not inconsistent with our experience of earlier revolutions in military affairs. For most – in one way or another – have impacted upon the warrior's view of his own profession. Take the use of the long bow (followed by the introduction of cannon) which destroyed the idea of chivalry, and brought into question the concept of active courage, or courage freely chosen. Courage became more passive in nature; it was valued in a new currency: blows received, not blows given.

The rise of mechanization on the industrialized battlefields of the late nineteenth and early twentieth centuries changed matters again by locking the warrior into a system in which his performance was increasingly evaluated in industrial terms: productivity and predictability. The warrior became a "worker." In our information age he has become an information processor, locked into a cybernetic world. And the biotech revolution promises to transform him again – perhaps, in more radical ways still.

The warrior meme

The biotechnological age we have entered prompts us to ask this question: Does a person's genetic makeup predispose him to various types of human behaviour? Is his chosen trade a preference or a biological disposition? Does it demand innate abilities or acquired skills? Is it a product of genes or memes?

Richard Dawkins popularized the idea that human development is memetic as well as genetic. The word "meme" was first coined in 1976 to describe the practices, conventions and taboos that codify genetic choices. Unlike a gene, which is copied to the offspring from its parents, a meme by analogy is anything that replicates itself or is copied via memory. Genes are instructions for making proteins, stored in cells of the body and passed on in reproduction. Their competition drives the evolution of the biological world. Memes, by contrast, are instructions for carrying out behaviour, stored in the memory and passed on by imitation. Their competition drives the evolution of consciousness and thus culture (Blackmore 1999: 30).

A meme can be anything from a good idea to a poem, anything, in other words, that spreads by imitation. In the case of the warrior ethos, the western meme was the example of the great warriors in *The Iliad*. Memes exist in a meme pool – like genes they evolve. They are copied, altered, and selected at different times. In the case of war we call this the evolution of the warrior ethos. Every era rewrote the Achillean legend for its own needs.

This was quite literally true in the case of Alexander the Great who carried the shield of Achilles 11,000 miles to India where it saved his life. Alexander, of course, thought that he was genetically descended from Achilles via his mother's line, but it was memetic repetition that was of historical importance. At the height of the battle of Granicus he turned the battlefield into a Homeric scene of single combat when he galloped out in front of his men and struck the Persian commander in the face with a spear. The fact that Homer's heroes were essentially armed with the same weapons as Alexander's hoplites made identifying with them all the more natural, and yet coming to terms with the phalanx and the unheroic infantry tactics of the day all the more difficult (Bartov 1996: 17).

Now, like genes, we must remember that memes adhere to the rules of Darwinian selection. They spread because they are good at spreading. The first rule of the replication of memes is that replication is not necessarily for the *good* of anything. In other words, there is no necessary connection between a meme's reproductive power (i.e., its fitness from its own point of view) and its contribution to our fitness. Yet whether we see the warrior as a hero or a parasite, the warrior meme was a powerful one, which is why it has endured for almost three millennia.

Today it can no longer be invested with the same weight and significance. For with the biotechnology revolution the old distinction between nature and nurture is becoming blurred. Genetics is becoming far more important than mimetics. Memes are giving way to genes. To put it crudely, if memes were designed (or programmed) to breed the warrior, genetic manipulation now promises the possibility of breeding out all the imperfections even the most heroic warriors have displayed through history. For the first time it offers the warrior a chance to *breed true*. Of the many technologies that are changing his sense of "self" three are essential to his – or her – "post human" future. The first is performative technologies. The phenomenology of human–machine interaction is changing as computers become more interactive and sophisticated. The task is to make us more machine-friendly, if not quite to see war from the machine's point of view. The second is in the behavioural arena. We have begun to turn the analytical tools of molecular biology into engineering tools. Most of the technologies today are compensatory (i.e., they compensate for injury or degradation of the body – spare parts/cosmetic surgery). Tomorrow, the human body will be enhanced through the fusion of organic and cybernetic material. The final arena is in the normative, where genetic manipulation and the use of synthetic drugs are rapidly extending the range of human actions beyond the possibilities enabled by natural selection. Both may also influence the way we think about ourselves in relation to others, especially our enemies.

Biotechnology and warfare – new soldiers?

In all three cases technology no longer involves an extension of the human body, as has been the case since the first tools and weapons were invented. It is now being incorporated or assimilated into the body at an increasingly fast pace.

Cybernetic warrior

To illustrate this, consider James Blinn's seminal novel of the 1991 Gulf War, a vivid portrait of a virtual warrior who finds himself taking part in history's first "virtual war." Blinn's dazzling book is the most evocative work on war since Joseph Heller's *Catch 22*, and while Heller captured what was essentially a Second World War soldier's response to what war was in the process of becoming, Blinn's is a forward-looking satire on what war has now become.

Set on a gigantic computerized aircraft carrier carrying a state-of-the-art arsenal bristling with "smart" weaponry to the Persian Gulf in history's first truly television war, the novel evokes what war now means to a twenty-first-century American. Asked what he does for a living the hero replies:

> I'm tempted to give her a dose of some acoustic techno-wizardry airborne and anti-submarine warfare jargon. Initiate a little battle of the jargonauts. Lay on some acronyms and abbreviations: ASW, FASOTRAGRU-PAC, ECS, MK-82, ADP, INCOS, SENSO, TACCO, COMNAVIRPAC, ECP, NATOPS, ESM, MAD, SAP, ACLS, AN/ALR 47, ASWWINGPAC. Or just gab along in navpubspeak: *The mission of the Sierra-3 Viking fixed wing, carrier based, all-weather, tactical anti-submarine warfare aircraft is to utilise its suite of active and passive computer-assisted detection sensors to localise track and terminally engage surface and sub-surface hostile contacts in electro-magnetically charged combat environments....* Instead I just say: "I'm in the navy."
>
> (Blinn 1997: 127–128)

Such is the authentic vernacular of the twenty-first-century warrior. If the language is technical this befits a soldier or pilot who occupies the same virtual space as his own weapons system.

Blinn's hero goes on to describe a world that he shares with a computer-machine, a world in which the traditional biological instruments such as the human gaze are now largely machine mediated. In a memorable scene he records how he saw a dead man for the first time with his own eyes, not through the usual medium of technology.

> I saw it with my own eyes. I saw it with my EYES.
>
> Therefore it's real. Isn't that how it goes? If you see it with your own eyes, in the same time and space as you're in, then it's real. If/then. Causality. That's it, right? It wasn't relayed or bounced or fibre-optic transmitted,

modulated or written or echoed, encoded, encrypted, or ciphered, projected through celluloid, optically etched on silver halide, simulated by ordering the polarities of magnetised ferrous atoms, facsimiled, thermal, laser or holographically imaged, analog or digitally processed, scan-converted, manipulated, synthesised, distorted, Animatronicized, equalised, morphed, tweaked, computer-enhanced, duped, dubbed, multiplexed or multitracked, photocopied, mineographed or inkjet printed, colorized, SurroundSounded, Dolby-ized, virtually-realised, or electron-energised on the back of a CRT ... No filters, no intermediary, no question of interpretation. Authentic, three-dimensional. Hard-wired. In my face. The stink of death.

(Blinn 1997: 261–262)

In both passages Blinn offers us a graphic picture of what war has become, and with it the warrior.

For the warrior meme has readapted to its environment again, as human and machine are increasingly assimilated. What we are witnessing is the interaction of man and machine, as both respond to the demands of a continuous stream of information. It is feedback that allows the system to function. The warrior's nature is *performative* because both organic (human life) and machine life work as one (cybernetically). Both exist by responding to external stimuli but they respond very differently than they did in the past (Graham 2002: 184).

Cybernetics is the science of a self-regulating system (one that corrects its own errors). A machine that is self-regulating and self-correcting is called a "servo-mechanism." Feedback – the use of information provided to determine the effectiveness of past and future actions – is essential because the system guides its own behaviour, using the results of its own past and present performance to determine its future performance. In this respect, it differs from Stimulus Response (SR), which determined the performance of soldiers in the past. In SR when a stimulus is presented, an organism (soldier) responds. A new stimulus presents itself and the organism responds in the same way at the same time. In both cases there was no inner processing, so much as an automatic response.

We can date the change to the work of Norbert Wiener, the father of cybernetics, who began work in 1941 on the firing mechanisms for anti-aircraft guns. Recognizing that if they aimed directly at a fast moving plane they would miss every time, he recognized that the only way to score a hit was to ensure that the guns were self-adjusting. Only by estimating where the targets were and how fast they were moving could they be targeted successfully. Information about their speed and direction had to be fed back into the aiming mechanism. As Wiener wrote, "To live effectively is to live with adequate information. Thus, communication and control belong to the essence of man's inner life even as they belong to life and society" (cited in Graham 2002: 185). The same can be said of the twenty-first-century warrior.

A modern fighter pilot is a technological breeding ground for a new kind of site-specific "self." This new subject position and subjectivity that operates under the name of "fighter pilot" is however not solely the product of its own unique institutional opportunities, specialised domains of knowledge and technology. It is also the product of a psychasthenic logic. For the fighter pilot's model body and its technologically articulated consciousness exist in their purist state at the site (the fighter plane's cockpit) of a fundamental disturbance of perception, that of a schizophrenic of self and place, the result of an organism's almost perfect assimilation to its surrounding space (Gray 1995: 260).

The man–machine interface, the new environment or computer space which machine and man inhabit together, is not an extension of the body, (like a tool) but a total environment. It is the context for a new corporeal reality, an entirely new world in which war is conducted, a world into which we are sensorially (not only physically) incorporated and assimilated.

But intelligence is still ontological, for it is still embedded consciousness that relies on biological factors such as neurological synapses and bioelectrical currents. Consciousness is an embedded phenomenon – the mind learns within a material environment. Computers are hard logic devices; only human beings have intuition and emotions that derive from embodiment. In war, as in life, spontaneity still prevails over programming. War is still an existential experience, though it is a different one from the world of Homer and the warriors of the past 1,000 years. For we are beginning to privilege the biological over the cultural, our genes over our memes.

Indeed, in a society in which soldiers are no longer encouraged to discover their self worth through the esteem of other men – or perhaps, more correctly are not allowed to, Western societies may only be able to engage in war not by emphasizing the cultural (the meme) over the biological (the gene) but by *emphasizing the biological at the expense of the cultural* – by re-engineering warriors through biotechnological means.

In civil society that trend is already in evidence. In-vitro fertilization may represent only one per cent of births in the US, and embryo selection a proportion that is not much larger. Cloning and genetic modification clearly lie in the future but the future can be glimpsed. We are escaping a humanity defined by natural selection. We are on the cusp of "overcoming" our humanity, to use a word favoured by Nietzsche. In determining our own evolutionary future we stand on the eve of a "post-human" age.

By drawing reproduction into a highly selective social process that is far more successful at spreading successful genes than sexual competition we truly are embarking on a new voyage. Within the next 50 years we will be able to modify ourselves, to design our own babies, and possibly better soldiers. The technological powers we used in the past to change the natural environment can now be directed at changing ourselves, by modifying, not so much human nature as the behaviour of specific human types, the warrior included. And this is made possible by breakthroughs in the matrix like arrays called DNA chips which may soon be able to read 60,000 genes at a time; the manufacture of artificial chromosomes which can now be divided as stably as their naturally occurring cousins; and advances in bio-informatics, the use of computer driven methodologies to decipher the human genome. The latter include old fashion steroids as well as new drugs that will enable us to enhance our physical capabilities (Stock 2003: 13).

Cyborg soldiers

The body is the first and most natural tool of man.

(Marcel Maus)

In terms of agency, soldiers/pilots have been inhabiting the same space with machines for some time now. So far the US military has been working on fusing the body and machine functionally, not meshing them physically. Systems analysis, social psychology, computer mediated systems and above all personnel management have all been designed to help pilots use machines more effectively in order to enhance human performance.

It is through cognitive engineering, however, that the US Air Force has gone further than any other service. For in modifying the cognitive processes of its pilots it has made them more efficient. In the case of the high performance computer based aircraft of today, pilots have to be capable of split second responses. Their minds have to be made more machine-friendly than ever. The goal of military training is now increasingly "design-oriented" – to design operators who can process information faster than ever (Gray 1995: 23).

Functionally, we are already wired into digital networks like the Internet that enhance our ability to process information. Some of the research projects already underway may be harbingers of the future. One of the most famous is the McDonnell Douglas' Pilot Associate Program, which has been ongoing since 1986. Designed to allow "expert systems" to evaluate the input from external sensors, as well as monitor and diagnose all the aircraft's on-board subsystems, including the pilot, it will be able to initiate actions if the pilot is unable to take decisions himself.

What is new, however, is that we are now exploring ways to mesh machine and body not only functionally, but also *physically*. Today a range of words and terms are now employed to describe our evolving cyborg status from biotelemetry to "human–machine interfaces" and bionics (the copying of natural systems). Engineering is being transformed into a biologically based discipline.

In the Massachusetts Institute of Technology (MIT)'s Artificial Intelligence Lab robots are assembled from silicon, steel and living cells. The activators of these simple devices are muscle cells cultivated in the laboratory, the precursors of the prostheses that will one day be installed seamlessly into disabled human bodies. Surgical body modification and biochemical alterations (for example through the use of botulinum toxin) are already commonplace. Within 50 years, or even earlier, these developments could be applied to enhance the abilities of tomorrow's warriors (*The Times* 2 February 2004).

One popular sci-fi example is the way that people interact with computers by incorporating silicon into their bodies. In William Gibson's cyber punk stories data is transferred via "wet-wired-brain implants," computer chips in their brains. These visions promise a world in which there will be a sophisticated interface between our nervous system and silicon; a world in which neural implants will enhance visual and auditory perception and interpretation, memory and reasoning; a world in which the distinction between computers and humans will be gradually elided.

Fibre optic projectors can already throw images onto our retinas, thus allowing us to see directly without a television or computer screen, and research is well underway to help enhance auditory senses through implants in the ear. And true to William Gibson's vision of the future the US Air Force is investigating growing neurons in silicon chips to improve the communication between human and machine – allowing chips to be activated by hormones and neural electrical stimulation. The Defense Advanced Research Projects Agency (DARPA) has a Brain–Machine Interface Program that, in its own words, aims "to create new technologies for augmenting human performance through the ability to access non-invasive codes in the brain in real time and integrate them into a peripheral device or systems operation." In plain English, this means enhancing human performance by working out how the brain controls movement and using it to control external devices – transmitting (as has been done successfully with monkeys) brain signals over the internet to operate a robotic arm hundreds of miles away. This experiment by DARPA may be the harbinger of an age in which a warrior's brain – now part carbon, part silicon – may be able to operate weapons by the power of thought (*The Times* 16 October 2003).

Cyborg warriors

"That there will be future warriors is the only certainty," claims Colonel Frederick Timmerman, Director of the Centre for Army Leadership and former Editor-in-Chief of *Military Review*. His is a future which belongs to those countries, principally, of course, the US which by transforming the way technology is applied will be able to "transform and extend the soldier's physiological capability" (Jordan 1999: 112–113). If war is to remain central to human culture then the soldiers' bodies, as well as personalities, may have to be reconfigured. In this regard, the cyborg condition has enormous implications for our humanity as well as our idea of war. For if endurance can be artificially enhanced will we have to rewrite the ethos of the warrior, as man and machine continue to co-evolve equally?

We must be quick to recognize, nonetheless, that cyborgs are not quite what sci-fi writers would like us to imagine – at least, not yet. The popular view of the military cyborg can be found in two recent Hollywood films, *Robocop* and *Universal Soldier*. In the first it is a product of the Detroit Police Department, a subsidiary of OmniConsumer Products (OSP). The "Unisols" (another industrial brand name) who appear in Robert Emmerich's 1992 movie *Universal Soldier* are cyborgs of a different stamp. Their hyper-accelerating bodies turn dead flesh into living tissue. Following their death in Vietnam, the two principal combatants in the film, are flown back home to be packaged in ice, surgically eviscerated and refilled with cybernetic equipment, and thereby transformed into true twenty-first-century soldiers. A serum injected into the back of their skulls voids their memories. They are the ultimate killing machines with no sense of fear in the face of death largely because, to all intents and purposes, they are already dead to themselves.

What both films offer the viewer is a vision of a future in which technology, pioneered by the private sector, not the state (the present RMA relies heavily on dual-use technology), is used by private companies as well. That too is one aspect of a larger picture – how technology is beginning to transform the ontology of war, as we have traditionally understood it. It is important to recognize that this technology can take three forms. It can be restorative – restoring lost functions, replacing lost limbs and organs. It can be reconfigurative – creating post human possibilities by adapting Man to his environment. The original work of Manfred Clynes (who first coined the term "cyborg") was on how to adapt Man to outer space. Finally, it can enhance human abilities, which is the aim of most military research.

Certainly, when it comes to restorative functions much progress has been made. But it is the third option that is probably central to the future of war especially if we can escape Darwinian evolution.

In re-engineering themselves will soldiers come to see themselves as members of an exclusive caste? The process of "technologizing" – in which bodies are reassembled so that they can function more optimally (i.e., so that excellence can be enhanced) is central to the cyberpunk sci-fi of William Gibson. In his imaginary world cyborgs are creatures whose identities are no longer determined by social criteria such as class, ethnicity or even nationality but by technicity (the new architecture of the body). In Gibson's world cryogenic processes and enhanced digitalized senses redefine identity, just as cyberspace produces its own virtual communities. In "Johnny Mnemonic" one of the principal characters has electronically upgraded vision and prosthetized fingers that house a set of razor sharp, double-edged scalpels, myo-electrically wired into her enhanced nervous system. She is no longer an individual (born into a social or ethnic group from which she derives her sense of self; she is a customized, functional product of a cyborg culture, and she has little respect for others who are not like her. What Gibson offers us is a vision of a separate caste – a world in which the respect one warrior has traditionally given another is no longer a product of culture, but bioengineering. What his cyborgs admire in each other is technical virtuosity and operational speed, attributes that have been directly integrated into their own prosthetic and genetic architecture (Gray 1995: 180).

Bio-warriors

The Human Genome Project represents one of the most significant steps in our continued evolution. On one level it allows us to subject our humanity which we have taken as a given to biotechnological intervention. In theory, it might be possible to breed a warrior DNA, or manufacture a race of warriors, or "natural born" killers. This is the promise of novels like *Ender's Game*. The object would be twofold – first, to make soldiers impervious to fear, flight or anxiety, and thus to make them more courageous (or foolhardy) in battle. The second is to make them more effective at killing. In pursuing the first, of course, one can accomplish the second.

For the "born" warrior is a killer, as well as one who is prepared – if necessary – to lay his life on the line. Joanne Bourke (1991: 1) writes that "the characteristic act of men at war is not dying, it is killing," and it is as well to always keep that in the forefront of one's mind. The soldier-killers analysed by Glenn Gray in his seminal study *The Warriors* (cited in Evans and Ryan 2000: 44) are among the most formidable and terrifying warriors of all – men devoid of remorse or reflection, and they exist in all armies at all times. Achilles was the supreme killing machine; so was the Alexander depicted so vividly in Arian's history of Alexander's campaign. Killing, Arian tells us without a trace of irony, is what Alexander did consummately well.

Is this a result of culture or nature? Evolutionary psychology argues that humans are born with a common set of preferences, predispositions and abilities fashioned by natural selection. These abilities enabled us to develop and become the dominant species on the planet. In other words, when we are born we are pre-equipped with abilities such as the ability to learn a language. We have something akin to a piece of computer programming that is designed specifically to enable us to acquire language skills. Experience – in the form of hearing our parents talk, for example – is just the input for the program.

It would also appear, however, that while we are all programmed to be violent, some individuals are genetically more prone to violence than others. One of the reasons that, even in the mid-twenty-first century war, is likely to remain a male activity is that, across cultures, men kill other men 20 or 40 times more often than women kill other women. And the great majority of killers are of an age in which the soldier is at his prime, usually between 15–30. It is also true, however, that in this same age group some are more inclined to kill than others. In Western society, for example, seven per cent of young men commit 79 per cent of repeated violent offences. If this is true of society in general it must surely be true of the military, which in a democracy at least, tends to be a microcosm of society as a whole (Pinker 2002: 131).

Of course, violence and war are not the same. A good soldier is not the personality type that makes an intractable violent young offender: impulsive, hyperactive, with a low intelligence and usually an attention deficit as well. Young offenders are also resistant to discipline, they dislike being controlled. And the worst of them are often psychopaths who lack a conscience and are more likely, if they join up at all, to be found in paramilitary organizations that set their own rules. Very rarely are they to be found in military units whose members are bound by close fraternal ties and tend, as a result, to enjoy a high degree of self-esteem.

But one of the interesting phenomena of war is the extent to which a very small percentage even of professional soldiers actually kill with any real enthusiasm. It has been calculated, for example, that one per cent of fighter pilots accounted for at least 35 per cent of enemy aircraft kills in the Second World War. Clearly, they were not only more talented but more aggressive than their counterparts. And on the ground the figures are even more remarkable. Take for example, Lance Sergeant Simo Hyha of the Finnish Army, who in three months in the Winter War of 1939 killed 219 Soviet soldiers with a standard issue service rifle. Another born killer was Sergeant Alvin York of the American Expeditionary Force in the First World War who killed 28 Germans in the battle of the Argonne on 8 October 1918, a month before the Armistice. Looked at differently, York single-handedly accounted for the equivalent of two German companies, for he also captured another 182, and as a result could have transformed a tactical situation along a key sector of the front line (Harvey 2002: 61).

Killing, in short, does not seem to come naturally in all situations to all soldiers, even the most highly trained. Natural-born soldiers are very few and far between. And that is why the military has preferred the discipline of collective units such as gun crews which are more easily controlled, and which being often distant from the battlefield are also less emotionally involved – in a word, more "mechanical." The greatest cruelties have been the impersonal ones of remote decision, system and routine especially when they can be justified as operational necessities.

So, looking at the situation from the perspective of the evolutionary psychologist, we can sum up as follows. We are not born to kill anymore than we are born to engage in war. Killing is a contingent strategy connected to complicated circuitry that allows us to compute subconsciously whether it is in our interest to kill, or not. We deploy aggression as a strategy and we do much less so than we have done before. Like war, violence has declined among the rich nations, not because of morality or ethical codes. Ethics tends to legitimize, after the fact, the contingent strategies we have already chosen. It is cultural and learned. If we have become more ethical (in our own eyes, at least) it is because we have become more cosmopolitan (or less hostile to strangers). And the technologies that promote literacy, travel and knowledge of history have all contributed in different ways to a growing cosmopolitanism. Our social imagination has expanded as a result. Through television and film we are able to project ourselves into the daily lives of other people, even though they are remote from us both physically and sometimes emotionally.

Of course, if our postmodern societies continue to discourage violence or sublimate it through sport then they may find themselves with a smaller pool of talent from which to draw natural born soldiers. And thus we may have to manipulate the gene pool to stay in the war business. Drug enhancement rather than genetic engineering is likely to be the key. For it is probably far too early to talk of the genetic engineering of soldiers, something, which if it happens at all, we are more likely to encounter beyond the half century.

One way of doing this, made popular by science fiction writers, is cloning – the transplanting of a mature human cell with its full DNA pattern into another human egg whose nucleus has been removed. Cloning is the way to transmit the genetic signature of one parent to an embryo and thus to create effectively a genetic identical twin of the parents. And this has given rise to the fear of "phalanxes of identical little Hitlers goose-stepping to the same genetic drum" (Nussbaum 1998: 55). But current opinion is that the cloning of human beings will remain genetically difficult, if not impossible for years to come. Instead, cloning is more likely to be used to provide cell banks for the living, i.e., replacing parts lost on the battlefield. The future of cloning probably lies in spare part surgery rather than the replication of human beings.

A more likely route is manipulation of our genes. By drawing reproduction into a highly selective social process that is far more successful at spreading successful genes than sexual competition we truly are embarking on a new voyage. Within the next 50 years we will be able to modify ourselves, to design our own babies, and possibly better soldiers. The technological powers we used in the past to change the natural environment can now be directed at changing ourselves, by modifying, not so much human nature as the behaviour of specific human types, the warrior included.

Already, we can modify a trait in a directed fashion by altering or selecting particular gene variants. Changing a single gene in an animal is now routine. And the research has been spurred on by scientists and their claim that they can decipher the relationship between our genes and our behaviour: such as criminality, alcoholism, and drug addiction. The trick is to identify a combination of gene variants common to many people with similar endowments (such as athletic prowess), and then to manipulate the human germline.

With so much genetic information available on every human being from simple single gene disorders to complex polygenic moods and behaviour traits it is becoming attractive for employers to use genetic data to select prospective employees. As early as the 1970s the discovery of the sickle cell anaemia trait prompted the US military to use genetic screening for the first time. Carriers of the recessive gene – most of them African-Americans – were denied entrance into the US Air Force Academy for fear that they might suffer the sickling of their red blood cells in a reduced oxygen environment (Bowring 2003: 212).

The US military purportedly went further in 1992 when it launched an ambitious programme to collect several million DNA samples from its personnel. The exercise was to enable the accurate identification of men and women lost in combat. But in the legal battle that followed the refusal of two marines to give blood under the Fourth Amendment right to privacy, the fear was expressed that the same genetic samples could be used for biomedical research to identify the best military genes, or to weed out soldiers with the worst: those most susceptible to fear (Rifkin 1998: 164). If it is possible to isolate genetic traits then it should also be possible to enhance personality traits like risk taking that would be required by Special Forces, and to produce above average levels of emotional stability for pilots in the virtual spaces that they occupy with computers.

Will this result eventually in the emergence of a warrior elite, a caste genetically distinct from other non-combat units? Traditionally the military has often seen itself as culturally distinct in its attachment to a value system that still honours courage, heroism, and indeed honour itself. But in a biotechnological age, culture may be transcended by biology. With the emergence of genetic screening why demarcate a warrior by class, or by ethnicity, or race? Why not do so on the basis of genotype – on the basis of positive discrimination: isolating certain "positive genes," or negative discrimination: screening to detect predispositions to mood and behavioural instability?

A much more profitable subject of speculation, because it has been happening for some time, involves the modification or control of human behaviour through neural-pharmacology. For in the immediate future the military will probably also try to manipulate the endogenous opiate system in an attempt to decrease sensitivity to pain, and thus enhance physical stamina and mental endurance. The genetically modified soldier is part of the Pentagon's search for an "Extended Performance Warfighter," a programme that focuses on using devices other than drugs to enhance performance. TMS or electro-magnetic energy may allow scientists to "zap" a soldier's brain giving him the capability to stay awake, fight and make decisions for a week. In the future, wearable devices attached to clothing may be also employed to gauge a soldier's mood by the number of eye blinks. Internal implants able to monitor the heartbeat may both be able to administer tranquillizers or sedatives without the soldier knowing.

The enhancement of athletic abilities by drugs has been with us in sport for a long time. Take the banned hormone Erythopoietin, which by raising the oxygen-carrying capacity of red blood cells can boost endurance by 10–15 per cent. Metabolic and physiological enhancers are now a central part of sport. The only question is their detection, and their side effects. The social pressure to re-engineer the athlete – for the athlete to win and to win more spectacularly than in the past – has generated the relentless use of pharmacology in sport. Likewise, the equally insistent need to win in war is likely to accelerate the use of military pharmacology.

Every army, of course, has attempted to reduce stress and thus improve combat performance by whatever means comes to hand: alcohol being the oldest. Today, through continuous advances in neuroscience it is proving cheaper, easier and much more productive to control anxiety and fear through pharmacological means. Librium and Valium already treat anxiety, Prozac and Zoloft depression. Prescribed drugs are regularly given to Air Force pilots to reduce stress, fatigue, and enhance wakefulness for up to 72 hours at a time. Anxiety suppressants are now given to pilots going into combat. Viagra is given to the Special Air Service (SAS) to boost testosterone levels and thus aggression.

But manipulating the emotions through pharmacology goes well beyond endurance or the reduction of stress. One day soon we may even be able to abolish guilt and thus neutralize the sometimes traumatic consequences of showing courage. What if by swallowing a pill a soldier could immunize himself from a lifetime of crushing remorse? For the prospect of a soul absolved by medication is about to become real. Feelings of guilt and regret travel neural pathways in a manner that mimics the tracings of ingrained fear; thus a way of addressing one should address the other. Experiments have been conducted at the University of California at Irvine to inhibit the brain's hormonal reactions to fear, softening the formation of memories and the emotions they evoke. The beta-blocker, Propranolol nips the effects of trauma in the bud. It is proving possible to short circuit the very wiring of primal fears (*The Times* 10 July 2003).

Another research team at Columbia University has discovered a gene behind a fear-inhibiting protein, uncovering the traditional "fight or flight" imperative at a molecular level. Will we soon be able to blunt the human conscience; to mediate out of the psyche regret, remorse, pain or guilt? Is this the ultimate end of "consequence management": making the soldier blind to the consequences of his or her own acts? Is this the thin end of a dangerous wedge: the emergence of a morally anaesthetized soldier?

Conclusion

Indeed, it is not in the existential but *metaphysical* dimension of war that the influence of biotechnology may be the most radical. The metaphysical is the way in which death becomes sacrifice, the way in which the death of a soldier has meaning for himself and others. Often that meaning is specific to a particular culture – it is not transcultural. Inevitably, ethical problems arise when communities have different preferences for ways of living, preferences which we can see as culture specific. All societies choose the best life for them – in the light of their own historical experience and collective self-understanding. It is at this point that ethical differences between communities arise.

For we all abide by universal codes, by the demands of human rights, which touch on an altogether different question: what the contemporary philosopher Jurgen Habermas (2003: 38) calls our "self-understanding as members of the same species." This concerns not culture, which is different in every age and every society, but the vision that different cultures have of humanity, which is the same. For Habermas, the biotechnological revolution threatens this self-understanding of the species. Indeed, he believes that recent developments in biotechnology and genetic research threaten to instrumentalize human nature according to instrumental preferences. The most obvious example is parents who want children of a certain skin colour, or hair colour, or parents who would like to breed out what they consider human imperfections, most of them genetic. The human body at this point is no longer sacred, because it becomes an "object," or an instrument of parents, or the state, to be modified or redesigned at will. And it is at this point, for Habermas, that our self-understanding as a species may be threatened. We may see human bodies merely as defective "hardware" and the mind as enhanced "software." Precisely the same might be said of the bioengineering of the warrior.

The ethical implications of biotechnology, therefore, are daunting. "The primary political and philosophical issue of the next century will be the definition of who we are" (Kurtzweil 1999: 2). In so far as the new technologies not only promise to remake our bodies but also our worlds they raise important and urgent questions about society's continued engagement with the soldiers who fight in its name. And in so far as war is likely to be a struggle between the West and non-Western peoples, states, societies or regimes its *inter-subjective* meaning has never been more important which is why even "post-human" warfare is likely to be just as ontologically real as before. In the words of one of America's leading contemporary philosophers, Richard Rorty (1999: 52):

> Humanity is neither an essence nor an end but a continuously and precarious process of becoming human, a process that entails the inescapable fact that our humanity is on loan from others, to precisely the extent that we acknowledge it in *them*.... Others will tell if we're humans and what that means.

It is the idea of humanity as a process that brings us to the core of the question of biotechnology and war.

In the future we will be encouraged to see humanity as a continuing process of "becoming" human, a process that through cyborg enhancement (a form of "participatory evolution") is now far more technologically determined than in the past. At the same time, morality is far more intersubjective than subjective. This is why the prospect that we may begin to fight "post-human" wars in the near future should prompt a sobering thought. Will tomorrow's western warriors find themselves alienated from a self-understanding of the same species? Will they think themselves genetically distinct from soldiers from other societies who are not experiencing the post human condition?

References

Bartov, Omar (1996) *Murder in our Midst: The Holocaust, Industrial Killing and Representation*, Oxford: Oxford University Press.

Blackmore, Susan (1999) *The Meme Machine*, Oxford: Oxford University Press.

Blinn, James (1997) *The Aardvark is Ready for War*, New York: Anchor.

Bourke, Joanna (1991) *An Intimate History of Killing: Face to Face Killing in the Twentieth Century*, London: Granta.

Bowring, Finn (2003) *Science, Seeds and Cyborgs: Biotechnology and the Application of Life*, London: Verso.

Card, Orson Scott (1991) *Ender's Game*, New York: Tom Doherty.

Evans, Michael and Ryan, Alan (eds) (2000) *The Human Face of Warfare: Killing, Fear and Chaos in Battle*, London: Allen & Unwin.

Frankowski, Leo (1999) *A Boy And His Tank*, New York: Simon & Schuster.

Graham, Elaine (2002) *Representations of the Post Human*, Manchester: Manchester University Press.

Gray, Chris Hables (ed.) (1995) *The Cyborg Handbook*, London: Routledge.

Habermas, Jurgen (2003) *The Future of Human Nature*, Cambridge: Polity.

Harvey, A.D. (2002) "Soldiers with a Special Flair," *RUSI Journal*, 147(1): 60–64.

Jordon, Tim (1999) *Cyberpower: The Culture and Politics of Cyberspace and the Internet*, London: Routledge.

Kurtzweil, Ray (1999) *The Age of Spiritual Machines*, London: Orion.

Metz, Steven (2000) *Armed Conflict in the Twenty-first Century: The Information Revolution and Post-modern Warfare*, US Army War College: Strategic Studies Institute.

Nussbaum, Martha (ed.) (1998) *Clones and Clones: Facts and Fantasies about Human Cloning*, New York: Norton & Co.

Pinker, Stephen (2002) *The Blank Slate: The Modern Denial of Human Nature*, London: Allen Lane.

Rifkin, James (1998) *Our Biotechnology Century: How Genetic Commerce will Change our Lives*, London: Orion.

Rorty, Richard (1999) *Philosophy and Social Hope*, London: Penguin.

Stock, Gregory (2003) *Redesigning Humans: Choosing our Children's Genes*, London: Profile.

The Times (2003a) "The No Mourning After Pill," 16 October.

The Times (2003b) "Are Cyborg Troops our Future Army," 10 July.

The Times (2004), 2 February.

3 Managing the revolution in military affairs

Ron Matthews

Blame it on Marshal Ogarkov, the Soviet General who, in the 1960s, first coined the term Revolution in Military Affairs (RMA) to describe the tumultuous changes occurring in the nature of warfare (FitzGerald 1994: 1). He thus ignited a debate that has raged for two decades. The debate among military historians and strategic studies experts focused on the underlying rationale for the revealed changes in the conduct of war. At the heart of the controversy lay the question as to whether these changes represented a seismic shift in the nature of war; that is, a fundamental discontinuity in the trend of weapons development and/or doctrinal thought, a veritable "revolution" in warfare, or whether the changes simply reflected a further cranking-up in the evolutionary development of battlefield capability (Matthews and Treddenick 2001). While the distinction between revolutionary and evolutionary warfare has essentially been driven by academic curiosity, the real work of planning and effecting military dominance has proceeded with a frenzy. Doctrine, policy, organizational structure, and weapon systems have necessarily had to change in response to the dynamics of the international strategic environment. It is for this reason that United States (US) policymakers dropped the grand, but contentious, RMA label, used to describe the present radical thinking on warfare, replacing it, instead, with the term, transformational warfare.

This chapter focuses on exploring, examining and evaluating the relationships, impacts and challenges associated with the dramatic change process unfolding. For consistency, the term RMA is employed in the discussions that follow, not least because it signifies dramatic and radical change, exactly what is happening in the defense context. To begin, the chapter will seek to identify the principal environmental drivers shaping the RMA. In this regard, the threats stemming from the emerging and parallel Revolutions in Strategic 'Affairs' (RSA), Technology (RTA) and Business (RBA) Affairs will be considered. These revolutions are multi-dimensional and complex; they are also driven by contradictions and paradoxes. It is unsurprising, therefore, that while achievement of the RMA has become a compelling goal for the world's defense communities, the military, political and economic viability of the concept is increasingly called into question. Having identified and examined the drivers pressurizing defense ministries to design and deploy policies to engage in the

RMA process, the chapter will then steer discussion towards the affordability of pursuing RMA goals, a topic of no little significance to small countries, with limited defense capabilities. Affordability has become a dominant issue in the RMA debate, due in no small measure to the existence of taut defense budgets and the escalating acquisition costs of modern weapon systems.

Decomposing value for money

Discussion will focus on affordability by reference to the policy position of the United Kingdom (UK) Ministry of Defence (MoD). Offering a case study of UK practices is justified by the fact that its MoD has been, and continues to be, at the frontier of policy developments to secure RMA affordability.

The overriding policy objective has always been the achievement of value for money (VfM). It permeates all aspects of UK defense management thinking and in practical terms is defined by the requirement that all defense-related expenditure on projects, products and services be judged against a public sector comparator (PSC).[1] The analytical structure of the cost-effectiveness relation, driving VfM, is illustrated in Figure 3.1. It seeks to explain the management and military revolutions that have been occurring simultaneously in the twenty-first century. A popular characterization of this change process focuses on the merging of the business and battlespaces. There is an RBA and an RMA, with the latter driving the former. The RMA is a representation of the effectiveness aspect in the cost-effectiveness relation. Effectiveness is defined as battle-winning capability, comprising a myriad of considerations, but principally the technical quality of weapon systems, doctrine and morale. By contrast, the cost element within the RBA focuses on the smart management of defense resources, pursuing policies

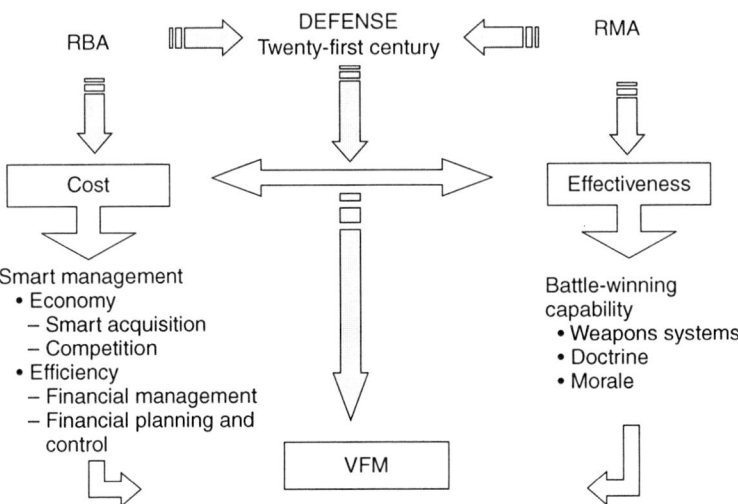

Figure 3.1 Integrating the business and battle spaces.

targeted on achieving both economy and efficiency of resource usage. The need to drive down cost through commercial solutions is premised primarily on economic and financial policy grounds.

Broadly interpreted, then, the search for VfM represents the search for economy, efficiency and effectiveness. The logic of the VfM policy framework wrests on the imperative of securing cost savings through the "smart" management of scarce defense resources, particularly those at the "support-end" of military endeavor, and relocating them to the "teeth-end" so as to maintain, if not enhance, military capability in the face of taut or even declining defense budgets. The effectiveness aspect (battle-winning capability) of VfM will not be dealt with in this chapter, aside from contextual comment. Instead, analysis will concentrate on economy and efficiency, and the linked policies to achieve affordability; but before evaluating the policies adopted under these two broad headings, it will be helpful if their contextual backgrounds are explained.

Introduction of "economy" policies have been linked to broader reforms in the management of defense technology. Since the launch of the UK MoD's 1998 Strategic Defence Initiative and the implementation shortly thereafter of the Smart Procurement Initiative, the pressure has been on to achieve ... "faster, better and cheaper" acquisition (Ministry of Defence 2004b: 2). The policy thrust was on cost reduction, as was the parallel "lean" logistics policy introduced in the early years of the millennium. Cost reduction, through an emphasis on economy, has become the fundamental theme of UK. defense management policy. Smart Acquisition and Lean Logistics are just two change management programs embroidered into the growing patchwork of policies, including competitive tendering, contractorization, arms trade liberalization, international arms collaboration and defense globalization.

Policies associated with the efficiency component of VfM focus, by contrast, on the imperative to raise output per unit of input, i.e., "more bang for the buck," or alternatively, the "same bang for less bucks." Central to efficiency reforms in a commercializing environment is the need to identify value. In the absence of accurate asset and liability valuations, it is clearly impossible to manage resources efficiently. Thus, as the UK MoD sought to increasingly commercialize defense, injecting greater transparency, accountability and business logic into decision-making, financial reforms have been introduced over recent years to enhance the level of management control. Of central significance in this regard has been the introduction of Resource Accounting and Budgeting (RAB). This facilitated the introduction of commercial accounts into the defense domain, enabling greater financial awareness and thus a more informed and efficient resource management process. Once established, RAB encouraged a cluster of derivative financial policies to emerge, including the Private Finance Initiative and the (Defense) Balanced Scorecard – a performance monitoring mechanism borrowed from the commercial sector.

The purpose of addressing the economy and efficiency components of VfM is to screen thinking on the rationale and application of the broad range of policies relevant to achieving affordability. However, beyond this, the chapter also seeks to evaluate whether such policies are appropriate in the defense context. The dilemma, of course, is whether the commercial and globalization revolution is encouraging acquisition from cheaper offshore vendors, particularly those in the US, undermining local defense-industrial sovereignty. However, in the changed strategic environment of the twenty-first century, is sovereignty any longer relevant? Perhaps, the RMA, reflective as it is of increased international interdependence, signifies the demise of national self-reliance as coalition warfare and global defense-industrial consolidation changes irrevocably the strategic map. Unquestionably, there is a need to balance strategic and affordability pressures. In the twenty-first century, the inescapable fact is that cost does matter.

The RMA – environmental drivers

The nature of weapons technology started to change during the late 1970s. Paradoxically, in a sense, the end of the Cold War accelerated the development and deployment of technologically sophisticated weapons and related command and control systems. Although technology has driven the change process, *ex post*, several distinct but nevertheless interrelated components of this RMA can be identified. Foremost, perhaps, was the changing nature of technology and doctrine. Gone was the era of Cold War superpower strategic standoff, of Mutually Assured Destruction, and the certainty of total war as opposed to limited conflict. The frightening specter of a Cold War between two superpowers had given way to an equally unappetizing prospect; that being, the proliferation of hot wars, the emergence of asymmetrical conflict (allied to a growing terrorist threat), and the enduring and pervasive sense of uncertainty. Capitalism had defeated the politico-economic threat of Communism, but it now faced, and continues to face, the less well-articulated threat of religious fanaticism.

Thus, against this strategic backcloth, the RMA matured to focus on the development of weapon systems to support expeditionary forces acting as a "force for good." War would henceforth be clinical and, indeed, economical, in terms of loss of life and collateral damage. Large standing armies, as previously maintained on the central European front, would now be redundant. They would be replaced with rapid reaction capability, emphasizing speed of deployment, mobility and lightly equipped forces possessing weapons enhanced by technology multipliers. The need was to proactively take the conflict abroad, rooting out terrorist havens and sanctuaries. For this strategy to be effective, however, "reliable" intelligence became an urgent requirement. Terrorist cells have to be identified and taken-out before deadly assaults on Western targets can be carried out. As a consequence, ISTAR (Intelligence, Surveillance, Target Acquisition and Reconnaissance) has emerged as a military priority. Technologies to effect ISTAR capability highlight the need for investment into the critical dual-use technologies of telecommunications, sensors, microelectronics, satellites, microwaves and neurotechnology. Digitization of the battlespace through C4I systems (command, control, communication, computerization and intelligence) is now viewed as critical to remove the "fog of war." Moreover, the introduction of standoff, laser-guided precision munitions means that surgical strikes against enemy positions have become feasible, reducing collateral damage to civilian life and property. Waging war on this basis benefits the US-led coalition forces on two counts. First, the lower level of innocent civilian casualties suffered in the countries harboring terrorists acts to win the hearts and minds of the local populace, reinforcing the view that coalition forces are a force for good. Second, the lower level of military casualties suffered by coalition forces acts to maintain public and political support at home.

Reinterpretation of national security

The appalling Al-Qaeda attack on New York's Twin Towers acted as the catalyst for intensifying the West's war against terrorism. The US focus on Homeland Security and the UK emphasis on the parallel concept of Resilience have sought to raise the profile of national security not only through the promulgation of domestic policies but also through the development of global security postures, policies and networks. Expanding the borders of national security recognizes the multi-dimensional foundations of terrorism; its roots, popularity and hence calling. While globalization may have facilitated the terrorist threat, there is a growing sense that globalization may also represent a partial panacea to the threat by ensuring that disaffected societies are embraced by the new emerging global socio-economic order. Attacking the basis of terrorism may require the West to refocus policy-efforts away from democratization, leading ultimately to development, by instead reversing the casual flow and promoting development as the basis for democratization (Boot 2004). The provision of bread to hungry people is arguably a more powerful force for good than electoral franchisement (Painter 2003; Gresser 2004). The obvious implication from such reasoning is that national security needs broader interpretation, extending beyond the boundaries of defense capability.

Globalizing the threat to national security

Although "security" has always meant different things to different people, a trend is emerging whereby interpretation of this amorphous concept is becoming more, rather than less, expansive. The broadening of the security focus is a reflection of the increased unpredictability of the post-Cold War international environment. The threat no longer derives solely from traditional interstate rivalry, but instead emanates from a wide spectrum of non-state sources. A broadening of the parameters of national security is all the more pressing since asymmetrical conflict embraces a whole cluster of non-conventional threats.

Moises Naim (2003) calls these non-state threats the "five wars of globalization"; i.e., illegal trading in drugs, weapons, people, money and intellectual property. Drawing attention to the proliferation of these non-state threats, a recent report of the United Nations (UN) High-Level Panel on Threats, Challenges and Change identified six clusters of threats to international peace and security, namely: transnational organized crime; terrorism; nuclear, radiological, chemical and biological weapons; internal violence, including civil war, state collapse and genocide; the continued possibility of interstate conflict and rivalry; and economic and social threats (Annan 2004; Russell 2004).

Attempts to define the RMA have become muddied not least by the ever-increasing fluidity of the threat but also by the apparent inadequacies of RMA doctrines and technologies. This is evidenced by the ongoing events of the Iraq conflict, where the concept of coalition warfare has been undermined by the absence of political cohesion; the reality, instead, being one of "coalitions of the unwilling." Infringements of the Geneva Convention, for instance, face the certainty of media transmission to a global audience, undermining international confidence in the processes and goals of military action. Equally, suicide bombings have caused unplanned and morale-sapping increases in the death and casualty rates of coalition forces, impacting negatively on public consensus and unity back home. However, whilst transformational weapon systems did prove successful in prosecuting the initial 2003 war against Iraqi forces, collateral civilian deaths and infrastructural damage have become a constant in the post-war equation. It appears that RMA technologies are impotent in achieving battlefield dominance against a determined and fanatical non-conventional foe. Moreover, damaging public relations revelations about the non-availability of basic equipment, such as body and tank armor for coalition forces, has led to a re-ordering of US defense budgetary priorities. This is symbolized by the January 2005 Pentagon announcement that future spending is to be concentrated on low-technology military acquisition at the cost of expenditure on high tech RMA-type technologies, such as F-22 Raptor fighters and stealthy DD(X) warships (Kirkpatrick and Pugh 1983: 16).

Cost, consolidation and capability

The other two principal drivers of the RMA are grounded in defense economics. The first has regard to the post-Cold War pressures on defense–industrial restructuring. These pressures came about after the ending of the Cold War, which left an international strategic vacuum, and removed the rationale for massive inventories of military platforms. This, combined with the recent static and even declining trend in global defense budgets, has led to excess capacity in arms production. The knock-on effect of market and government pressures has been to push defense companies towards consolidation. The 1992 "Last Supper" heralded the US consolidation process, with the 1999 Letter of Intent taking Europe down the same path. Pressures to consolidate have been driven by commercial and often political imperatives to reduce duplication.

The second defense economic driver influencing the nature of the RMA also had its origins in the post-Cold War strategic environment. Changes in the nature of war and defense doctrine focused policy and production on technology-based capability instead of platform volumes. The ensuing reorientation of defense industrial output encouraged increased integration with commercial information, telecommunication and computerized technologies and systems. Enhancement of weapons-related technologies have become exceedingly expensive, however; and the problem is that increases in defense budgets have not kept pace with rising acquisition cost. As a result, structural disarmament has occurred, whereby numbers of weapon systems acquired continuously decrease through each generation of fighter, or other complex weapons system. Kirkpatrick (1995) illustrates this rising unit cost by reference to consecutive generations of military aircraft. His research indicates that acquisition cost is escalating by around 10 percent per annum, much higher, of course, than the increase in most countries' defense expenditure.

Weapons acquisition cost has risen to such an extent that even the advanced countries find the required expenditure prohibitive. Unsurprisingly, the trend has been away from national acquisition solutions towards off-the-shelf purchase (from global, mostly US vendors) and collaborative procurement; the former often involves licensed production of existing foreign weapon systems, whilst the latter tends towards consortium approaches to both development and production of new weapon systems. High acquisition cost has meant that, for instance, UK national programs for fighter aircraft no longer exist, with land and naval systems also looking vulnerable to global acquisition solutions. Moreover, the UK MoD intends that international arms collaboration will be the acquisition vehicle for 40 percent of all Britain's future acquisition requirements. At the regional level, the multi-national Tornado and Typhoon programs are examples of high profile collaborative acquisition approaches in the European sphere; at the global level, there is the US-dominated F-35 Joint Strike Fighter. The latter is a truly international consortium comprising the US, UK, Australia, Canada, Denmark, Italy, the Netherlands, Norway and Turkey. There are several tiers defining the nature and extent of the partnership, with the US as the lead country. The UK, as a first tier partner, has invested US$2 billion into the program, representing approximately 12 percent of the project, but enjoying around 20 percent of the work-share. F-35 work packages can only be awarded on a competitive basis, and Britain with its technologically advanced and highly diversified defense-industrial base, is thus well positioned to exploit F-35 manufacturing opportunities. Smaller countries, such as Norway, the Netherlands and even Australia, are not so well placed. Indeed, with around 90 percent of the F-35 work already allocated, some of the partner countries have already begun to vent their frustrations (Matthews 1992).

Regional and global arms production consortia have proved an attractive acquisition option for countries seeking to realize their RMA ambitions. Technology sharing is an obvious way not only to leverage expensive R&D through partnership but also to capture economies of scale via integration of customer markets. The experiences of Tornado and Typhoon consortia appear to validate the view that multi-national collaboration is cheaper than national acquisition, but this is a highly controversial area of defense economics. Theory and folklore suggest that while there is a cost premium on collaborative production, this is outweighed by considerable savings on R&D (not least because the latter is shared amongst several participant nations) (Matthews 2001: 73–79). The production cost premium arises because each of the country partners insists on possessing separate assembly facilities. This means that instead of one European factory complex enjoying scale economies from specializing in the production and assembly of all 650 Typhoons, scale is fragmented; the four partner countries each assembling the aircraft that they intend to purchase (Spiegel 2005).

Aside from arms collaboration and related technology-sharing initiatives, defense companies have sought to defray high acquisition costs by exploiting globalization opportunities. As with commercial practices, defense companies operate in highly competitive global markets, and to maintain or gain competitive edge, they have actively outsourced components and sub-assembly manufacture to low-cost overseas sites. Additionally, defense companies have forcibly had to engage in "offsetting" defense production to arms-purchasing countries. Irrespective of the involuntary nature of this process, the end-result is one that often results in the permanent insertion of offshore subcontractors into global defense-industrial supply chains.

In spite of continuous pressures to consolidate, collaborate and globalize over the last two decades, the incessant cost escalation of high technology RMA weapon systems has meant that severe affordability problems still remain. The UK MoD, for example, has been suffering budgetary pressures almost continuously since 2000. In January 2005, it was forced to delay the signing of the £800 million Watchkeeper program to build unmanned reconnaissance drones because of intense pressures on the weapons procurement budget (Spiegel 2005). Significantly, delays have occurred due to the insufficiency of MoD funding to complete the required R&D. This delay comes as other larger defense projects have run into similar funding constraints. Indeed, nearly every large UK defense acquisition program, not currently under contract, faces hold-ups due to budgetary pressures (Chuter 2005). These include the Royal Navy's two multi-billion pound next-generation aircraft carriers (Spiegel 2005), the RAF's £13 billion fleet of air-to-air refueling tankers and the army's £6 billion program to build a new generation of armored vehicles (*Defense Acquisition* 2002). Given the cost escalation and budgetary challenges facing the UK, and indeed, other defense ministries worldwide, it was inevitable that the search for affordability would become a policy priority.

The parallel revolution in business affairs

The logic behind the RBA is thus the need to ensure affordability. Policy measures can be devised to target critical cost areas across the defense budget, with priority areas almost always embracing acquisition and logistics. For the UK MoD, the more efficient management of defense technology has become an absolute priority. Introduction of the Smart Acquisition policy was intended to lead to 10 percent savings across a ten-year period, and an even more ambitious savings target for lean logistics of 20 percent over five years (Ministry of

Defence 2004: 4). While claims have been made that dramatic savings have been achieved, caution should be exercised in the interpretation of such savings. Questions need to be asked, including: what is the baseline for calculating such savings? Have they come about through increases in efficiency or are they simply a reflection of cost reduction through service deletion? Furthermore, tellingly, where will future savings come from? This is important because savings will become progressively more difficult to secure. However, notwithstanding such questions, Smart Acquisition has become the central plank of UK acquisition reform.

Economy

Smart acquisition

The concept of smart acquisition grew out of the 1998 Smart Procurement Initiative; the latter originating in the earlier Strategic Defense Review. The purpose of Smart Acquisition is to improve efficiency of the procurement process. The UK MoD commissioned McKinsey, a global consultancy group, to design the policy framework for acquisition reform. The changes were indeed dramatic. A new CADMID (Concept, Assessment, Demonstration, Manufacture, Inservice, Disposal) cycle was introduced, replacing the old Downey cycle (Harrison 2004). The number of decision gateways were reduced from four to two, with the intention of speeding-up the decision-making process. Upfront target investment in the Assessment phase was recommended to be at least 15 percent of the acquisition budget (Spiegel 2004a). The aim of this recommendation was to ensure design robustness at an early stage of the cycle, thereby reducing both technical and financial risk later in the program. Whilst meritworthy as a policy goal, in practice it would always be challenging to achieve. Reality, of course, dictates that resources would be prioritized for the maintenance of ongoing defense programs rather than investments anticipated to yield results 10–20 years into the future. Examples of programs starved of upfront development expenditure are not difficult to find. For example, the troubled Astute program achieved just 0.8 percent of its £3.5 billion budget in upfront project expenditure; worse still, the equally calamitous Nimrod program took a miniscule 0.1 percent of its £3.6 billion budget for Assessment investment, pre-Main Gate (Speigel 2004b).

A principal tenet of Smart Acquisition is the emphasis given to partnership. Accordingly, Integrated Project Management Teams (IPTs) were established, comprising representation from the principal stakeholders, such as MoD, defense industry, the relevant branch of the Armed Forces, QinetiQ, as well as asset management organizations. A cooperative approach to acquisition would be encouraged, replacing the adversarial model employed in earlier times. IPTs would remain in existence throughout the project's life-cycle, from concept to disposal. Thus, even though constituent personnel would change, there would still be continuity of project oversight and ownership. IPTs would be empowered and customer-focused, arguably highlighting the imperatives of leadership and customer-need.

A major impact of the Smart Acquisition process has been to raise the level of debate and awareness of the challenges policymakers face in securing greater VfM, yet real and sustainable progress in achieving "faster, cheaper and better" defense acquisition remains sporadic and challenging. A recent National Audit Office (NAO) Report (2004) stated that for 2003–04 the UK's 20 major defense projects incurred additional costs of £1.7 billion and slipped a further three months behind schedule. This compounded the £3.1 billion extra costs and added slippage of 18 months recorded for 2002–03. Added to the previous cost overruns, the total cost of the UK's biggest defense acquisitions, accounting for about two-thirds of overall procurement had reached £50 billion higher than initial budget estimates (Spiegel 2004b). Aside from the continuing inability to achieve "faster" and "cheaper" acquisition, two worrying developments emerge from the NAO Report. First, there now appears to be little difference between new and inherited programs. Strikingly, the NAO Report argues that this split is no longer a relevant distinction because, as analysis shows, many so-called "smart" projects have failed to apply smart acquisition principles consistently. Second, the project displaying the biggest cost increase is the F-35 Joint Strike Fighter (JSF), supposedly the latest and best example of an RMA–RBA program (Odell 2003).

An important feature of these smart management reforms is the increased emphasis on competition. Since the 1980's Levine era, competitive tendering has been held by the MoD as a superior form of contractual process for achieving VfM, particularly cost reduction. The prevalence of cost-plus pricing as a consequence has fallen, and is now used solely for first-of-series acquisitions where future costs are unknown. In these non-competitive non-risk contracts, progress milestones are used to monitor program performance and activate payments to contractors. Where contract risk can be more easily predicted the award of fixed price contracts has become the norm. Nevertheless, problems may persist. The reality is that fixed price often does not mean fixed price, and certainly does not absolve the MoD from picking up the tab when contractual terms unravel. The MoD, for example, was forced to bear an additional £700 million costs of the £1.45 billion refinancing package for the troubled high technology Nimrod and Astute programs (Matthews and Parker 1999: 34).

"Costs" of competition

The theory underpinning competitive tendering is unremarkable and non-controversial. However, as with much management theory, it is the process of implementation that causes difficulty. Competitive tendering is presumed to reduce cost when compared to cost-plus contractual arrangements. However, it is difficult to obtain evidence to support this argument, not least because of the counter-factual nature of the comparison; that is, once the competitive tendering option is pursued, then a cost-plus comparator simply does not exist. There is a view, however, that the MoD is able to make rule-of-thumb estimates regarding the benefits of competition, possibly by comparing the reduction in contract price with the costs incurred in running the competition (Levene 2000: viii). The imponderable is the value given to contract price reduction. It arises from the difference in the price secured through competitive tendering and that which a monopoly supplier might be expected to charge. Yet, it is unlikely that the latter value can be estimated precisely.

There are claims that the MoD's policy of competitive tendering has led to cost savings of up to £1 billion per year, or a collective 10 percent saving, but doubt has been expressed over the veracity of such figures (Kemp 2004: 11). Schofield (1995: 148), for instance, is unconvinced, arguing that neither the claims for savings nor their relationship with competition can be adequately tested. Calculating the benefits of competition is even more challenging if the costs of competition are factored into the equation. The costs are several. First, the competitive tendering process often delays the decision to award a contract. The original Bowman contract, for example, is a classic case where the "illogicality" of a limited number of firms bidding against each other precipitated the collapse of the competition through competitors joining forces to submit a unified bid. As a consequence, the MoD was obliged to stop the competition, restarting it at a later date. Moreover, delays in contract award for typically complex defense competitions are endemic in the UK. The Royal Navy's multi-billion pound carrier competition is mired in what appears to be an unending series of contractual disputes (Milner 2005). The initial competition for choosing the prime contractor was protracted, with the MoD "failing" to pick a winner, deciding, instead, to plumb for shared prime contractor responsibility. Tensions ensued over the nature of the design, build and integration partnerships, exacerbated by the inability of stakeholders to agree contract price and carrier size. Initially, fears were raised as to whether the 60,000-ton warships were too big to access any of Britain's naval dockyards. More recently, the F-35 fighters destined to fly from the carriers were reported to be over-size and over-budget (Ministry of Defence 1995: xxxi). Finally, there has been wrangling over the remit given to the physical integrator. The role of this latter company is of critical importance within the MoD's alliancing strategy for the carriers' acquisition. However, the integrator's role is worth just £5 million in a program likely to be at least £3 billion. BAe Systems, in particular, is seeking clarification as to the integrating company's management control. Given the delays and cost overruns on its Nimrod reconnaissance aircraft and Astute submarine that have blighted the reputation of BAe Systems and its relationship with the MoD, Britain's biggest defense contractor is rightly nervous that if problems arise with the carrier program, it will suffer the blame even though it does not have the responsibility.

If competition delays do occur, then extra costs may be incurred by MoD having to retain obsolete equipment in service beyond the planned replacement date. The competitive tendering process may also suffer from what might be termed over-competition. This occurs when excessive numbers of firms engage in the competition, causing industrial and MoD scrutiny costs to rise pro rata with the numbers of bidding firms. Historically, there have been as many as 15 firms bidding for UK defense contracts, though the average number of bidding firms now hovers around six. A National Audit Office (NAO) Report (1994) indicates that across 13,000 competitions, the number of companies invited to tender averaged only 6.4. Arguably, however, even six bidders is excessive, leading to considerable costs of competition. Other criticisms of the contemporary process of competitive tendering include the frequency of competitions, over-detailed nature of requirements, excessive use of subcontract competition, exposure of proprietary design information to competitors – both domestic and foreign, and the transfer of excessive risk to contractors through elongation of the competitive tendering process. The bidding process is inherently a costly business as normally only one bid will be successful. Ultimately, the cost of bidding falls on all stakeholders, including both those firms submitting nugatory bids as well as the MoD, which as a consequence means an increased burden for taxpayers.

The NAO estimates that bidding costs amount to about 3 percent of contract values, yet this may be conservative as there is evidence to suggest that tendering companies, in order to be compliant with bidding requirements, spend up to 5 percent of contract value to stand any chance of success in the competition (Matthews and Parker 1999: 35). The scale of these competitive costs can be shown by reference to absolute values. For instance, a detailed and compliant tender for even a relatively simple piece of equipment such as a military vehicle costs between £500,000 to £1 million (Bell 2000: 33). However, bidding costs obviously rise, *pari passu*, with increases in the complexity of a weapon system, and the Maritime Patrol Aircraft is a case in point. Its bidding and assessment costs have been quoted as exceeding £100 million, and much of this is arguably wasted expenditure (Odell and Eaglesham 2003). However, such huge competition costs are not an isolated event, and are beginning to occur with some frequency. The competition process for Britain's carriers reportedly also cost the MoD £100 million, with £30 million allocated to each of Thales and BAE Systems for funding their respective bids and £40 million allocated for the MoD scrutiny process (National Audit Office 1994). Moreover, the bid process is ongoing and the build-date continues to slip.

The costs of competition can represent a sizeable percentage of the total acquisition cost, but this begs the question as to whether the acquisition cost can be estimated with certainty (National Audit Office 2004). Two issues in particular obfuscate the mathematical precision characterizing the cost estimating process. First, there is the problem of "entryism." This is where competitors (focused on down-selection) and MoD acquisition teams (focused on program approval) are under pressure to reduce cost and time estimates, to secure, arguably, short-term goals at the expense of long-term objectives. The 20 biggest UK defense acquisition programs are currently 14 percent above costs forecast at approval, raising concerns as to the quality of cost estimates underpinning the bid process (National Audit Office 2004). The second factor impacting on the quality of acquisition cost estimates is more directly connected to the RMA. The fact is that the duration and complexity of developing contemporary Smart Weapons and associated systems makes it unlikely that requirements will remain constant throughout the early phases of the CADMID cycle. In the twenty-first century RMA era, advancing technology, evolving threats and changing doctrines increase the likelihood of significant systems changes. Defense contractors anticipate the near-inevitability of such design changes, and may be incentivized to win work through "low-ball" bidding, being prepared to subsequently face the challenges of defending margins.

Competitive tendering and Smart Acquisition policies have been applied, with mixed results, to achieve greater economy in the management of scarce defense resources. Supplementing these cost reduction policies have been the introduction of efficiency reforms targeted particularly on financial management. This has required a raft of financial management reforms designed to improve the identification, transparency and communication of MoD financial activity. The "Department" is vast, and, accordingly, there is a recognition that cost control across the MoD needs to be improved.

Efficiency and financial management

Resource accounting and budgeting

RAB is defined as commercial accounting, based on the accruals principle. The accruals approach recognizes creditors and debtors and associated accounting conventions which were totally absent in the MoD's former cash-based accounting methodology. RAB allows defense resources to be valued, thus facilitating a more efficient management approach to be adopted. Critically, RAB, through the identification of opportunity costs, allows the true cost of defense to society to be calculated. New Zealand provides a classic example of a country that introduced RAB policies, enabling it for the first time to make informed cost-effectiveness judgments. Partially as a consequence of these fresh financial insights, New Zealand's MoD took the radical step of disbanding its air force, based on the grounds that it was not cost-effective.

The MoD annually publishes a set of RAB accounts, comprising the following statements: Operating Cost Statement; Balance Sheet; Cash Flow Statement; and two Productivity (input–output) Statements, linking identifiable operational aims with the budgeted cost of their provision. The accounts have several notable features. First, the Operating Cost Statement, as its title suggests, focuses on MoD costs rather than the revenues and profits that would be generated and thus reflected in the accounts of a commercial organization. When MoD revenues are earned (for instance, from disposals of surplus military equipment) they are treated as negative costs in the operating cost accounts.

The Balance Sheet lists the values of the UK's defense estates and assets, including all warships, main battle tanks and fighters. The balance sheet contains a number of important features. First, short- and long-term liabilities will be deducted from the asset total to provide a net asset value; that is, the financial worth of UK defense capability. Second, long-term assets, including weapon systems, listed in the Balance Sheet will be depreciated. These assets are shown at their current annual book value, net of depreciation – reflecting the cost of the asset's consumption over the last financial year. The annual value of depreciation moves to the operating cost statement, adding to the cost of UK defense. A third feature of the Balance Sheet is the "cost of capital." The cost of capital is calculated as a percentage of net assets and also moves to the operating cost statement as an expense. As stated earlier, the MoD's Operating Cost Statement shows annual cost accumulation rather than profit determination. Both the cost of capital and depreciation are legitimate, though notional costs of providing defense capability. In the commercial sector, accounting regulations require companies to identify all costs, including interest on capital and depreciation, before charging these costs against revenue to record profit. The lower the costs recorded the higher the benefit (or profit) to shareholders. By the same token, the MoD is now seeking through RAB to identify all costs, including the cost of capital and depreciation, to determine the true burden of defense borne by UK taxpayers.

The Cash Flow Statement is the third financial report within the MoD's RAB accounts. The statement provides a breakdown of the defense budget in cash terms, forming the basis for parliamentary debate and the vote for the budget's acceptance or rejection. Parliament's vote is for a sum of money, budgeted to be spent on actual goods, services or military activities. Thus, not included in this cash total will be the "notional" items of depreciation (representing movements of funds within an organization) and the cost of capital (where, in reality, public sector budgetary allocations derive from costless tax revenue).

The final two sets of RAB accounts are Productivity Statements, highlighting the budgeted costs of achieving Departmental operational objectives. These input–output statements facilitate evaluation of the MoD's financial efficiency policies; the aim being to achieve a given level of output from reduced inputs.

Resource accounting represents a major financial innovation across the MoD as well as the broader UK public sector. Through resource valuation, RAB has afforded the introduction of sensible and responsible resource management decision-making. Importantly, RAB illustrates the economics of choice and associated trade-offs. For example, the technique allows valuation of the high opportunity-costs of maintaining high and often obsolete war stocks making the case for introducing "just-in-time" ordnance supply systems. Equally, RAB highlights the high cost of keeping the Procurement Agency in London instead of relocating it to lower-cost areas of the UK and provides awareness of the strikingly high costs of consultancy provision (£549 million in the 2003–04 RAB accounts) compared to the costs of financing the Royal Marines (£604 million in the same year) (Timmins 2005).

Capital sourcing

Alongside RAB, there has been revolutionary change in the MoD's approach to sourcing capital for major defense projects. No longer does the MoD automatically provide capital for high-value investments; instead, the policy is to promote partnership under the overarching rubric of Public Private Partnerships (PPP). PPPs seek to replace public with private debt, embracing a wide spectrum of capital sourcing possibilities. These include contractorization, outsourcing, operating leases and perhaps the most important of all, Private Finance Initiatives (PFI). A PFI is where a consortium of commercial organizations is established, comprising companies in, for instance, the construction, asset management and engineering fields, to finance the building and servicing of a defense capability over a specified period of time. Examples of defense PFIs include the £500 million, 30-year Joint Services Command and Staff College project, the refurbishment of MoD Main Building, Whitehall, and the planned £13 billion Air-to-Air Aircraft refueling program.

As at March 2005, 633 PFI projects with a capital value of £40 billion had been signed (Ministry of Defence 2004). The purpose of PFIs is to reduce the cost of capital projects by the leveraging of private sector enterprise, expertise and commercial efficiency. Nearly all the defense PFI projects are in the support area, but increasingly those closer to the front-end of military operations are being considered as PFI contracts. The Defense Procurement Agency now has to demonstrate that a PFI is not the best financing option before embarking on any other procurement strategy. However, as always the qualifying condition is the attainment of VfM. The criterion in this regard is that the PFI contract costs less than the Public Sector Comparator; that is, the cost of in-house MoD provision.

It will not be possible to gauge the success of PFI contracts until the contractual period expires, and for most contracts this will take up to three decades. However, evidence is emerging that attests to the value of the PFI model. A Treasury study, for example, indicates that 85 percent of PFI contracts have been delivered on time, with 80 percent coming in on budget (Draper 2004). The MoD states that PFI contracts on average cost 10 percent less than conventional contracts (Ministry of Defence 2004a: 149). Whilst the charges required to service these contracts is anticipated to reach 3.5–7 percent of the defense budget by 2008, the MoD argues that they would have generated capital savings of up to £4 billion.

Financial planning and control

Achievement of VfM pervades all defense management activities, and no more so than through the MoD's relationship with the Treasury. To incentivize performance, the Treasury imposes efficiency targets, such as manning rates and efficiency savings. These targets are linked to the control and monitoring aspects of the MoD's performance management framework, particularly the Defence Management Board (DMB). This latter body, the equivalent of a commercial company's Board of Directors, is entrusted with monthly evaluation of MoD performance. The DMB uses a "management-by-exception" methodology, within a bespoke Defense Balance Scorecard methodology, to identify and then resolve weak performance wherever it emerges across the Department.

The role of Financial Planning is now an important element in the MoD's new financial management regime. Defense budgeting is devolved from top-level budget holders down through lower tiers of budgetary responsibility, with transparency and accountability at every stage. It is a far cry from the unscientific business of determining defense estimates in the 1970s and 1980s. Then, the MoD simply submitted estimates to the Treasury. These estimates were perused and invariably agreed; there was no control, just monitoring. Moreover, it was a rigid budgeting system. Any MoD underspend was lost; any overspend gave rise to a penalty. The problem with this approach is that MoD projects typically have lives of 20–30 years, and invariably do not go to plan; they carry high levels of risk, are complex, prone to delay, and involve considerable and often uncertain through-life costs. Under this old budgeting regime, if an acquisition program was efficient and went at a pace faster than budget, then funds might not be available, and other budgets such as training may have to be cut to finance the acquisition spend. Due to the inflexibility and inbuilt inefficiency of this financial system, it was inevitable that reform would come with the advent of Smart Management techniques.

The reforms came in 1998, with the introduction of a new Comprehensive Spending Review (CSR). Out went the old rigid annularity system, and in came flexibility, with the MoD able to carry forward budgetary underspends. The budget is now agreed for three years ahead; the base-year for the next cycle remaining unchanged from the final year of the previous cycle. Biennial reviews then determine expenditure for the next two years. On a monthly basis, the MoD reports to the Treasury, providing a detailed breakdown of actual defense spending against figures in the CSR. In order to recalibrate resource inputs and move towards departmental outputs/outcomes/targets, the 1998 CSR required that budgets would be tied to Public Service Agreements (PSA). The PSAs derive from the aim a particular Government Department sets itself. Thus, for the MoD, its aim is to: "Deliver security for the people of the United Kingdom and the Overseas Territories by defending them, including against terrorism; and to act as a force for good by strengthening international peace and stability" (Ministry of Defence 2003). Within this aim, PSA targets include the need to: achieve success in military tasks undertaken at home and abroad, including those providing support to civil communities; deliver the equipment program to cost and time; and strengthen European security through an enlarged and modernized North Atlantic Treaty Organization (NATO) and enhanced European defense capabilities (joint target with the Foreign and Commonwealth Office). These, of course, are "big picture" PSAs and will be reinforced at the micro-level by more specific targets, such as raising operational readiness levels or improving recruitment and retention rates by a pre-determined percentage. The requirement to achieve PSA objectives against budgeted defense expenditure represents yet further refinement to financial management policies designed to secure efficiency and affordability in the defense context.

Conclusion

This chapter has attempted to highlight the increasingly important relationship between the economy and efficiency resource management inputs and the effectiveness of defense capability outputs. Two revolutions are thus occurring simultaneously, one in business affairs and the other in military affairs. The former supports the latter in ensuring the affordability of transformational warfighting capability. In critical evaluation of these relationships, a number of the principal commercial and financial reforms introduced into the UK defense sector have been profiled to establish the nature and extent of RBA-type policies. Defense is an expensive business, and for most countries represents a sizeable proportion of their public spending. Government therefore has an overriding responsibility to achieve VfM from its spending programs. By definition, this requires that economy, efficiency and effectiveness are maximized. If managing the introduction of RBA reforms has become an imperative for advanced countries, like the UK, the policy implications for smaller, less developed, countries are even more profound. Application of rigorous defense management policies allow informed decisions to be made. At the strategic level, it may mean the abandonment of a particular military capability, as in New Zealand's decision to disband its Air Force. Equally, at the micro-level, while policies such as stock deletion appear less dramatic, the cumulative effect of these and other benefits accrued from parallel VfM initiatives, can make a substantial contribution to achieving defense affordability. Given the increasingly high cost of weapons acquisition, affordability is the challenge now faced by all defense ministries across the globe. Other small armed forces will unlikely be immune from this challenge.

Note

1 PFI deals are premised on the fact that project costs from private sector financing will be lower compared to conventional public sector arrangements. Thus, to justify PFI financing, it needs to beat the net present value of the conventional (MoD) public sector contractual model, incorporating whole-life costs. See "Realising the Potential of PPP," *Private Finance Journal*, vol. 4, no. 5, November–December 1999.

References

Annan, Kofi (2004) "Courage to Fulfil our Responsibilities," *The Economist* 4 December 2004.

Bell, M. (2000) "Leaving Portsoken: Defence Procurement in the 1980s and 1990s," *RUSI Journal* 145(4): 20–36.

Boot, Max (2004) "The Spread of Freedom Benefits Rich and Poor," *Financial Times* 16 September 2004.

Chuter, Andrew (2005) "UK JSFs Delayed, Carrier May be as Well," *Defense News*, 7 March 2005, p. 3.

Draper, C. (2004) *DBSA Generic Induction Briefing*, Director General SMART Acquisition.

FitzGerald, Mary C. (1994) "The New Revolution in Military Affairs," *RUSI – Whitehall Paper Series*, London.

Gresser, Edward (2004) "Counter Terrorism with more Trade Ties," *The Straits Times* 13 July 2004.

Harrison, Michael (2004) "MPs Attack MoD's £3 billion of Cost Overruns," *Independent* 21 October 2004.

Kemp, Damian (2004) "UK Battles with Procurement Challenges," *Jane's Defense Weekly* 2 June 2004.

Kirkpatrick, David (1995) "Starship Enterprise Revisited – Prospects for the 21st Century," *The Hawk Journal* (RAF Staff College) 1995 issue: 20–29.

Kirkpatrick, David and Pugh, P. (1983) "Towards the Starship Enterprise – are the Current Trends in Defense Unit Costs Inexorable?" *Aerospace*: 16–23.

Levene, Lord (2000) "Letters: The Levene Reforms – Some Replies," *RUSI Journal* 145(5): vii–viii.

Matthews, Ron (1992) *European Arms Collaboration*, Durham: Harwood Academic Press.

Matthews, Ron (2001) "International Arms Collaboration: The Case of Eurofighter," *International Journal of Aerospace Management* 1(1): 73–79.

Matthews, Ron and Parker, Judith (1999) "Prime Contracting in Major Defense Contracts," *Defense Analysis* 15(1): 27–42.

Matthews, Ron and Treddenick, John (2001) *Managing the Revolution in Military Affairs*, Cambridge: Palgrave.

Milner, Mark (2005) "BAe Says Carriers Face MoD Wreck," *Guardian* 31 January 2005.

Ministry of Defence (1995) *UK Defence and Trade and Industry Committee Report* HC61/62, London: HMSO.

Ministry of Defence (2003) *Delivering Security in a Changing World*, Defence White Paper.

Ministry of Defence (2004a) *MoD Annual Reports and Accounts 2003–04*, London: HMSO.

Ministry of Defence (2004b) *Smart Acquisition Handbook* 5th edn, London: HMSO.

Naim, Moises (2003) "Five Wars of Globalization," *Foreign Policy*, January/February: 28–37.

National Audit Office (1994) *Ministry of Defence: Defence Procurement in the 1990s*, Report by the Comptroller and Auditor General, HC390, London: HMSO.

National Audit Office (2004) *Ministry of Defence: Major Projects Report 2004*, Report by the Comptroller and Auditor General, HC 1159 – I session 2003–04, London: HMSO.

Odell, Mark (2003) "Whitehall Re-draws Key BAe Defence Contracts," *Financial Times* 20 February 2003.

Odell, Mark and Eaglesham, Jean (2003) "Carrier Bidding has Cost £100m," *Financial Times* 28 January 2003.

Painter, Anthony (2003) "Can Developing Nations be Independent?" *New Statesman* 17 November 2003.

Russell, Alec (2004) "Carrier and Jets to Go in Bush Defence Cutbacks," *Daily Telegraph* 31 December 2004.

Schofield, S (1995) "The Levene Reforms: An Evaluation," *Defense Analysis* 11(2): 147–174.

Spiegel, Peter (2004a) "Agency Chief Sets his Sights on the Bottom Line," *Financial Times* 10 November 2004.

Spiegel, Peter (2004b) "MoD Warned on Delays and Cost Overruns," *Financial Times* 10 November 2004.

Spiegel, Peter (2005) "MoD Postpones £800m Contract to Build Army's Scout Drones," *The Financial Times* 26 January 2005.

Timmins, Nicholas (2005) "PFI Information to be Posted Online," *Financial Times* 9 March 2005.

Part II

Transformations and operations

4 The essentials of effects based operations

Joshua Ho

From the start, it was clear that Operation Iraqi Freedom was going to be a different kind of war, one that would be characterized, in the words of Tommy Franks, Commander Central Command (CENTCOM), "by shock, by surprise, by flexibility, by the employment of precise munitions on a scale never seen before and by the application of overwhelming force" (Beck and Downing 2003: 105–6). Franks' bold claims were borne out: the time taken to secure victory was short for a major campaign, precision weapons were used more extensively than in any other conflict, and the coalition casualties sustained were relatively low for a major operation. In contrast to Operation Desert Storm 12 years earlier, Operation Iraqi Freedom took half as long and required only one-third as many troops to complete the mission (Noonan 2003).

Although the above statistics are impressive, some have suggested that the United States (US) was able to win because Iraq was a "paper tiger" with a military possessing obsolete equipment and little desire to fight (Beck and Downing 2003: 120). In addition, the no-fly zones imposed after the first Gulf War gave the US a tremendous advantage in keeping the Iraqi air force in check and facilitating the conduct of surveillance prior to the start of the war. Some have even alleged that the actual war began as early as December 2002, when suppression of enemy air defense (SEAD) operations purportedly began (Cordesman 2003). Nevertheless, the ground battles – at the Al-Faw Peninsula, Umm Qasr, Nasiriya and Basra and eventually the move into Baghdad – were neither bloodless nor walkovers (Beck and Downing 2003: 18–38), and victory was certainly never predetermined. Notwithstanding these difficulties, the swift and overwhelming nature of the coalition victory has prompted many to claim the arrival and coming of age of a new way of war known as effects based operations (EBO). The publicity about EBO as a new way of war after the successful conduct of Operation Iraqi Freedom may have shed more heat than light on the topic; still, several theories of what an EBO is can be discerned from the existing literature. This chapter examines the capabilities required to conduct EBO, the extent of their use in military operations conducted since 1990 with particular reference to Operation Iraqi Freedom, and the capabilities and limitations of these capabilities to provide evidence of the current state of practice of this way of war.

Key principle of effects based operations

Shaping behavior through effects

EBO seeks to move away from a destruction-centric, attrition-based and linear approach to warfare; rather, it seeks to separate the means from the ends by identifying the outcomes or strategic objectives desired in a campaign and then deriving the means required to achieve those outcomes. Implicit in EBO is the focus on shaping behavior of adversaries to such an extent that they will choose to surrender and not fight if possible. The means or "effectors" can comprise both military as well as non-military instruments, such as psychological operations, diplomatic, economic and political methods. The breakthrough in thinking comes in recognizing that destruction is not an end in itself but a means to an end. This way of thinking also points to how the impact of physical action is felt not only in the physical domain but also in the cognitive or psychological domain of the adversary. Another approach recognizes that every action has immediate effects, also known as primary effects, secondary effects, as well as tertiary and unintended effects, and treats the adversary as a complex adaptive system (Gellmann 1994: 17).

Certainly the concept of EBO is not new. The emphasis on behavioral change echoes the strategic thought of Sun Tzu (1987) and Carl von Clausewitz (1976), both of whom emphasized the importance of psychological aspects of war, to the extent that action in battle may be to gain psychological advantage alone rather than to achieve a physical objective. More recently, Liddell Hart (1972: 338–9) asserted that the aim of strategy was to dislocate the enemy, which would result in either dissolution or easier disruption of the enemy in battle. Douhet (1942: 20) believed that a successful air attack on civilian infrastructure deep in enemy territory would create tremendous moral and material effects on the civilian population and break the enemy's national resolve to fight. As early as 1941, the principles of EBO provided the doctrinal underpinning for the Air War Planning Document No. 1 of the US Army Air Corps (Mann *et al.* 2001: 93), which provided the doctrinal basis for subsequent campaigns such as the air raids on critical nodes in the German war economy such as the ball bearing factories in Schweinfurt, the destruction of which would have knock-on effects on the German war economy's ability to maintain production of war materiel (Gleeson 2001: 28–9).

These historical antecedents notwithstanding, the dominant approach to warfare then was still focused on the destruction of the adversary's military assets. After Operation Desert Storm, there was a rethink of how the US should go about fighting future wars in a post-Cold War environment of unequivocal American military dominance (Morley 2003). Following the collapse of the Berlin Wall, there were pressures within the US Department of Defense (DoD) to shrink military budgets, reduce the number of capital assets and the number of military personnel, leading eventually to a 40 percent reduction in the total deployable force (Cordesman 2003: 14). Given this background, the concept of

EBO was initially used to consider how future operations could be conducted with leaner force structures. Subsequently, the thinking evolved into knowing how and where to apply precise force in order to achieve a rapid and decisive victory (Ullman 2003: 11).

Theories of effects based operations

Most commentaries on the employment of EBO in Operation Iraqi Freedom actually refer to the theory of rapid dominance. However, this is just one theory of EBO. Indeed, six different interpretations of EBO can be identified from the literature. One treats EBO as a planning methodology for the conduct of operations, which emphasizes the strategy-to-task links, the integration with other planning processes and the use of both military and non-military means to prosecute the adversary. Operation Iraqi Freedom provided one such example of this planning process: where the strategy (of conducting regime change, eliminating weapons of mass destruction and capturing the terrorists in Iraq) was integrated with the operational objectives; which were translated to: (1) finish the regime, (2) eliminate weapons of mass destruction, (3) capture or drive out terrorists, (4) gain intelligence on terrorists and weapons of mass destruction, (5) secure oil fields, (6) deliver humanitarian relief, (7) create the conditions for representative government, and (8) ensure territorial integrity (Beck and Downing 2003: 105). In turn these eight operational objectives were translated to specific missions of component commanders, which were in turn translated into tactical action plans. The strength of the planning process lay in the fact that each mission could be traced back to an operational objective.

The second theory treats EBO as efficient targeting (Deptula 2001). This approach seeks to exploit the key weak points of the enemy by analyzing its capabilities as a total system. It focuses on the destruction of critical nodes rather than the destruction of the entire infrastructure to achieve the desired effect, and on the conduct of parallel operations, which emphasizes simultaneous, not sequential, attacks on all desired targets. US Joint Forces Command uses the term effects based targeting to describe efficient targeting. Two examples illustrate how this concept is achieved in practice. In Desert Storm, in order to nullify Iraqi air defenses, the US decided to attack the two major sector operations centers (SOCs) providing command and control to the air defenses. Even though the specifications stipulated that six 2,000-pound laser-guided bombs were required to totally destroy the hardened SOC bunkers, the US chose partial destruction via a single 2,000-pound bomb which then "smoked out" survivors from the building. This allowed more aircraft sorties to be generated for strikes against four more SOCs discovered subsequently. In Iraqi Freedom, a less protected communications switch located 200-meters away was destroyed instead of the command and control bunker that was underneath Baghdad's Rashid Hotel, which at that time had a number of foreign journalists staying there. Attacking the communications switch that served the command and control bunker had the effect of rendering the bunker ineffective, as it could no longer communicate with the troops under its charge (Beck and Downing 2003: 107).

The third theory treats EBO as the application of all instruments of national power, political, military, economic and diplomatic, to nullify all elements of adversary national power (Gleeson 2001: 11–15). It claims that the reliance on a single source of national power will inevitably reduce the overall effectiveness of a campaign and make it relatively easy for an adversary to adapt to this single form of attack. Two examples will illustrate how this theory is achieved in practice. The air campaign conducted as part of Operation Allied Force against Serbia aimed at effecting a Serbian military withdrawal from Kosovo. Although all non-traditional techniques were employed, including the modeling of social networks to persuade those close to Milosevic to influence him to withdraw his troops from Kosovo, NATO's decision to forgo the threat of a ground invasion meant that the air campaign in Serbia did little to persuade Milosevic to change his policy towards Kosovo (Lambeth 2002: 13). It was the Kosovo Liberation Army ground offensive supported by the US which proved to be the deciding factor in Milosevic's withdrawal. However, the extensive damage to infrastructure resulting from the air campaign had the unintended consequence of eroding Milosevic's political support and led to his eventual fall from power. Operation Iraqi Freedom also provided a limited example of the use of this concept, at least until the end of the hot war phase. The United Nations (UN) Security Council had sought to use non-military instruments such as economic sanctions against Iraq's oil exports before resorting to the use of force. The gathering of a coalition for the operation and its execution using combined forces ensured that the US was not acting alone.

The fourth theory treats EBO as rapid dominance. It relies on the employment of a series of unrelenting "waves of powerful strikes across many targets combining sea, air, land and space forces to affect and influence the adversary's perception and includes the physical capture and occupation of territory if necessary" (Ullman and Wade 1998: 1–2). Operation Iraqi Freedom provided a visible example of this theory where cruise missile strikes and air bombardment were conducted on hundreds of targets in parallel with the deployment of maneuver forces on the ground. The maneuver force moved with such rapidity that they reached Baghdad airport, just 20 km from the city center in 13 days.

The fifth theory focuses on EBO as interaction and collaboration between the operational commander and the other key actors in a campaign in order to deal with uncertainty in operations arising from a complex and adaptive adversary (Gleeson 2002: 6–9). In concept, the interaction between the operational commander and his civilian leaders, his tactical commanders and the sources of knowledge about the enemy as complex adaptive systems are learning experiences for the participants not only in the planning but also in the execution phases of a campaign. Appealing though it may be, there is little indication that this theory was applied in the ground battle during Operation Iraqi Freedom, although a limited application was the dynamic incorporation of lessons learned into the coalition's execution of operations.

The sixth theory focuses on EBO as network-centric warfare. The theory considers the adversary as a complex adaptive system (Smith 2002: 26). It proposes that four ingredients are required to deal with complexity and uncertainty: the ability to generate different action options for decision-makers by linking diverse sets of engagement and sensor capabilities; the ability to adapt to an intelligent adversary's actions through shared awareness of an unfolding situation; the ability to coordinate complex actions in synergy at different levels through shared situation awareness and common understanding of command intent; and the ability to mobilize knowledge and expertise to bear to provide timely support to decision makers at all levels. The networking of resources in the entire war-fighting enterprise is proposed as the way in which to master uncertainty and deal with complexity (Alberts *et al.* 1999: 87–93). Although there was no demonstration of this networked form of warfare since the 4th Infantry Division, reputedly the most networked force in the US Army, was not used in Operation Iraqi Freedom, the dramatic results achieved in the US Army's Division 21 advanced war-fighting experiments – 75 percent of current combat platforms can defeat over twice the number of enemy forces, in half the time, at over three times the size of the battlespace when networked – may provide proof positive of the efficacy of network-centric warfare (Alberts *et al.* 1999: 180).

The practice of effects based operations

These differing interpretations on EBO notwithstanding, their proponents agree that several requisite capabilities are needed (Shanahan 2001: 2). Since EBO link physical action to effects, the abilities to conduct and integrate precise physical actions plus the skills to observe the effects arising from those physical actions are important capabilities to develop. In addition, in order to generate the next set of actions, there is a need to determine if the effects generated have met desired outcomes and are able to communicate both the information generated from effects assessment and the decisions for the next course of action. Capabilities to sense and understand developments in the battlefield, and to be able to communicate knowledge derived from battlefield information, and to create precise effects are thus needed.

Sensing

Sensing is the process of identifying targets to be struck, cuing action upon positive identification and evaluating the effects achieved from the action. Manned, unmanned aircraft and space-based platforms with its attendant sensors used during Operation Iraqi Freedom and sensing technologies employed in the war have advanced to such an extent that achieving positive identification for fixed installations and static weapons emplacements has become a reality.

Area-wide search and locate capabilities were provided by satellites, long-endurance high-altitude and medium-altitude unmanned aerial vehicles (UAV), as well as manned surveillance aircraft. The National Reconnaissance Office (NRO) employed three advanced KH-11 visible and infrared imaging spacecraft, and two to three Lacrosse all-weather imaging radar spacecraft that provided 24-hour coverage to image fixed installations, detect Iraqi armor, static weapons emplacements and missiles launches (Fulgham 2003: 22). Global Hawk, a high-altitude UAV, was used as a strike co-ordination and reconnaissance asset, and was especially effective in locating air defense and surface-to-surface missiles through its synthetic aperture radar (SAR) that could see through sandstorms (Streetly 2003). Medium-altitude UAV, like the Predator, were used in surveillance and autonomous strike. Manned surveillance assets such as the U-2 high altitude surveillance aircraft and the Joint Surveillance and Target Attack Radar System (JSTARS) were used extensively to provide dynamic surveillance and targeting during sandstorms.

Signals intelligence (SIGNIT) provided another valuable source of sensing information, obtained via electronic eavesdropping with the RC 135 Rivet Joint aircraft, which can be used to sift airwaves for mobile and cellular telephone transmissions and locate the caller's position. The capability proved particularly useful in locating surface-to-surface missiles and SAM launchers, as their operators gave their position away through too much chatter. The Iraqi leadership were also located when they were forced to use high-frequency radio, which was easily intercepted and exploited for intelligence once their fiber-optic landline and public switching networks were interdicted (Knights 2003).

The combination of sensing capabilities across usable bands of the electromagnetic spectrum on platforms operating at different altitudes allowed continuous surveillance and targeting to be performed under different anti-aircraft threat situations. However, there are three limitations to current sensing capabilities: the inability to detect and identify high-value, well-protected mobile missile launchers; the failure to discriminate less well protected mobile targets, like trucks, from civilian vehicles; and the incapacity to detect, identify and discriminate targets hiding in foliage or within buildings (Vick *et al.* 2001: 32–6, 40–3, 64–5, 110–15, 121–33). Future developments are likely to focus on solutions like the development of satellites with Ground Moving Target Indicating Radar (GMTI), Synthetic Aperture Radar (SAR) and Inverse Synthetic Aperture Radar (ISAR) capabilities, enhancing Global Hawk with GMTI features and foliage-penetrating radars, and developing mini- and micro-UAVs to detect and identify targets under foliage, and within or under infrastructure.

Managing knowledge

The aim of sensing is to collect information about the adversary and about the efficacy of previous actions. This sensing information provides the basis for anticipating the adversary's likely behavior, as well as one's own course of action. Both software models and associated hardware are needed in the generation of subsequent courses of action. The models assist in anticipating the effects of hitting specific targets so that sensor platforms can be appropriately positioned to monitor those effects. Equally crucial is the ability to convey the information from the sensors to the knowledge processors and communicate the courses of action to the players who will execute the missions. Both knowledge creation and knowledge communication are principal components of knowledge management.

Creating knowledge

Current linear targeting models center on John Warden's Five Rings theory of warfare, which contends that the war effort should be directed primarily at the enemy's physical side as the moral or human side is beyond the realm of predictability. In Warden's model, the critical core is the enemy leadership, while the orbiting systems are organic essentials like infrastructure, population and the opponent's fielded military forces (Shanahan 2001: 4). Latter adaptations include the National Elements of Value (NEV) model that details the relative importance of the target systems to the national leadership and the relative importance of the target systems to each other; the agent adaptation model, which determines the various ways an adversary might react to an attack based on three scenarios: what is most beneficial and what is most restrictive to the adversary; what is known about the adversary's capabilities; and what if the adversary had certain currently unknown capabilities (Fayette 2001). The Input–Output model takes each target as an input factor and the outputs are the effects of striking the targets on different sectors of the military, like active forces and reserves, level of training, weapons store, status of weapons, and logistics ability. The model allows one to determine the target that has the most effect on the elements of military power and assists in the analysis of possible secondary and tertiary effects.

The system currently being developed to enhance targeting is known as the Automated Assistance with Intelligence Preparation of the Battlespace (A2IPB), where soldiers input the latest battlefield conditions into the program and it replies with the enemy's most likely next geographical move for the Air Force and the Army. Terrain, weather and force organization information is used and matched with templates of enemy doctrine to predict the future movement of enemy forces (Ames 2003). Data mining to identify new connections between subjects, the use of artificial intelligence to help sift through information and correlate large volumes of information like satellite imagery and enhanced visualization systems to display the information and knowledge generated will be features of the system (Kenyon 2003). The A2IPB interoperates with target development systems, ISAR management and employment systems, fusion systems, and intelligence command and control databases.

The targeting models assume that the physical effects achieved will translate to behavioral outcomes. However, historically, this has not been shown to be true. Behavioral models address this weakness and incorporate both a targeting model and a situation-aware, recognition-primed (SARP) decision-making model to determine the required actions needed to shape adversary behavior. The recognition-primed model postulates that all decisions flow from analogies drawn from both the current and previous situations that have been experienced by a person. It asserts that a person frames the existing situation by recognizing the patterns from a previous experience, and matches that to the current situation. Subsequently action is taken based on the actions that the person has previously taken (Klein 1998). The SARP goes one step further by incorporating prospect theory into the model. Prospect theory allows one to determine an individual's propensity for risk and the kind of actions individuals with different risk profiles would undertake. By incorporating prospect theory, the model does not need a store of the adversaries' previous experiences, but seeks to affect their perception of the situation through alteration of their appetite for risk.

Despite the power of behavioral models like SARP, they fail to recognize the adversary as a complex adaptive system (CAS). A CAS is one in which the interacting autonomous and semi-autonomous entities comprising the system can adjust their behavior as a result of externalities acting on the system. CAS models incorporate targeting and behavioral models and include statistical and probabilistic methods to model the non-linearity of adversary behavior. CAS models are the most powerful of all the models. Because CAS models are so powerful, they require high-performance systems that are capable of self-learning to drive the model. Although advances in information technology systems suggest that such a self-learning system might be technologically feasible within the next few years, the development of a cultural–military–economic model is still lagging behind developments in information technology (Shanahan 2001: 8).

Communicating knowledge

Another aspect of hardware is the communications backbone needed to communicate command intent. In Operation Iraqi Freedom, satellite communications were used extensively to convey command intent and aid collaboration between commanders, superiors and peers in the fast-moving battlefield. So heavy was the requirement for bandwidth that commercial satellites were used to meet 84 percent of the requirements (Williamson 2003). The Global Command and Control System (GCCS), which can use satellite or radio frequencies for transmission, was also a critical backbone in providing accurate location of blue forces down to the platoon level. Based on the desired need for more bandwidth after Operation Iraqi Freedom, the Distributed Common Ground System (DCGS) enhances an architecture that is capable of integrating command, control, intelligence and surveillance operations across globally distributed forces (*PR Newswire* 2003). In addition, DCGS is also backward compatible and able to incorporate future modular change.

Effecting

The next step after sensing the environment, deciding on the course of action to take based on an analysis of adversary information, and creating knowledge through computer models is to create precise effects through physical action with "effectors."

Precision guided munitions

The first class of "effectors" is precision-guided munitions (PGMs) and there has been a noticeable trend of the increased use of PGMs in conflicts (Cook 2003). The increased usage can be attributed to the increasing accuracy of PGMs developed since World War II (Cerasini 2003: 11). During World War II, 1,500 B-17 bomber sorties were required to drop 9,000 bombs to destroy a target of 600-meter square in size (Crowder 2003: 16). During the Vietnam War, the accuracy of precision weaponry had improved to such an extent that the same 600-meter square target only required dropping 176 bombs from 30 F-4 sorties. During Desert Storm, the laser guided bombs proved so accurate that they accounted for 75 percent of the damage upon Iraqi strategic and operational targets, even though they constituted only 4.3 percent of the total tonnage expended. The technology had improved to such an extent that by the time of Operations Enduring Freedom and Iraqi Freedom, up to 24 similar targets could be targeted by one B-1 sortie with the GPS-guided Joint Direct Attack Munitions (JDAM) (Crowder 2003: 16).

Parallel improvements in stealth capabilities of aircraft have also allowed the bombing missions to be carried out more effectively. As vital installations and other high-value targets are often well protected by radar-guided guns and missiles, a force package of aircraft is usually assigned with the bombers to neutralize air defenses in order to get bomb-dropping aircraft in and out of the target area safely. During Operation Desert Storm, a force package of 33 aircraft required to protect eight bombers embarked on a bombing mission, translated to an escort-to-bomber ratio of about 5:1 (Crowder 2003: 17). By the time Operation Iraqi Freedom was conducted, the increased use of stealth aircraft meant that a bomber could proceed for a mission with literally no escort aircraft, that is, one F-117 sortie was able to deliver two bombs to just as many targets (Deptula 2001: 18).

Now that stealth and precision technologies have matured, the next level of development will be to make attacks more surgical through the development of small-diameter bombs of about 130-kg. These small-diameter bombs are more suitable against small, mobile targets and urban targets (Sirak 2003a). Other developments include combining precision with an all-weather capability by fixing laser seekers to the current JDAMs; increasing stand-off accuracy by attaching a wingkit to the JDAM (Sirak 2003c) and through air-to-surface missiles like JASSM (Sirak 2003b); and developing a responsive call-for-fire system with a loitering precision attack platform like the RQ-5A Hunter unmanned air vehicle (UAV) (Burger 2003b) or the modified Tactical Tomahawk (TacTom) missile (*International Defense Review* 2003). Hence, dramatic improvements in the accuracy of PGMs over the last 60 years and the parallel development in stealth technology have made it possible for the US to conduct strikes on infrastructure deep in enemy territory with limited collateral damage to civilian personnel and infrastructure, and the ability to facilitate the conduct of EBO.

Maneuver forces

Another tool used to create effects during Operation Iraqi Freedom was the deployment of mobile forces like armor and armored infantry. Although the air campaign did much to reduce Saddam's ability to command and largely reduced the combat power of the Iraqi Army, pockets of resistance by irregular forces still held out. The coalition knew that it had to insert forces into the capital quickly in order to force the regime to capitulate. 5th Corps bypassed urban areas and headed straight for the jugular, reaching within 50-miles of Baghdad in five days (*Army Magazine* 2003). The sight of M1 tanks and M2 Bradley fighting vehicles entering the capital was more than sufficient to convince the Iraqis that Saddam's regime was no longer in control. When the statue of Saddam Hussein was pulled down, this act provided the proverbial final straw that broke the regime's back. Future developments will likely focus on better support to facilitate speed in operations, either by improving the efficiency of the maneuver platforms or by improving the effectiveness of the logistics support.

Special forces

The employment of Special Forces was also a key feature in Operation Iraqi Freedom. Active mainly in the north and west of Iraq, Special Forces comprised nearly 8 percent of the combined force package and managed to narrow the battlespace from a California- to a Connecticut-sized space (Noonan 2003). In effect, it was Special Forces coupled with air power working in concert with the lightly armed local Kurds and the 173rd Airborne, which effectively replaced the 4th Infantry Division, and formed the Northern Front. Special Forces were also involved in the liaison with Kurdish forces to ensure that they took no action to prompt Turkey to invade (Burger 30 Apr 2003a). Special Forces proved to be so useful that they were assigned multiple roles, including directing air attacks and raids against a terrorist camp on the Iraqi–Iranian border, searching for Baath leadership along the highways from Baghdad to Tikrit, seizing selected targets like oilfields to prevent Iraqi leadership from setting them on fire, holding dams to prevent the leadership from flooding large parts of the country, and occupying airfields for subsequent use by the coalition and denial of its use by Iraqis who may have intended to launch Scud missiles at Israel.

They also held key towns in the north and important buildings like the presidential palace in denial missions, disrupted internal Iraqi lines of communication in Baghdad and other command and control facilities, as well as provided information on the whereabouts of Iraqi leaders, which ultimately aided attacks against Saddam Hussein and his spokesman, General Ali Hassan Majid ("Chemical Ali") (*Strategic Comments* 2003). There were also reports that the US military, the Central Intelligence Agency (CIA) and Iraqi exiles conducted a broad covert effort inside Iraq three months before the start of the war to forge alliances with Iraqi military leaders to persuade them to cooperate and not fight (Filkins and Jehl 2003).

The multi-role capability of Special Forces and its civilian equivalent, the CIA, was a highly desirable factor in EBO as they could fulfill and perform a variety of missions ranging from surgical destruction, psychological operations, persuasion, and liaison that contributed to the overall creation of effects. Future developments in this area will likely focus on further integration of Special Forces and regular force operations and the provision of on-call firepower to Special Forces.

Information operations

Another "effector" was the conduct of information operations to directly influence the psyche of the Iraqis. A psychological war was waged with over 50 million leaflets dropped over Iraq and hundreds of hours of radio/television broadcasts made to scare the Iraqis into inactivity or desertion. Many of the leaflets were dropped even before the war began. They contained instructions on how to surrender and gave warnings of the consequences for anyone thinking of using chemical or biological weapons. In addition, text messages were sent to the mobile phones of individual Iraqi commanders to persuade them not to fight. Jamming of communications nodes was another strategy used to neutralize the Iraqi air defense system without destroying them (Kenyon 2003). Other information operations included "communications herding," whereby most frequencies were jammed, forcing the Iraqis to broadcast from a small set of other frequencies that were more easily disrupted or exploited for intelligence (Koch 2003b).

Besides persuasion, deception was the flip side of information operations. Saddam was led to believe that the war would start later than it did by deceiving him into thinking that the 4th Infantry Division was a vital part of the war, even though it was not. This was achieved by keeping the 4th Infantry Division floating off Turkey after it was clear that they would not be allowed to transit through Turkey, and by sending troops of the 4th Infantry Division slowly to the Gulf to give the impression that the US needed to open the northern front in order to succeed. Both actions caused Saddam to leave the oilfields in the south relatively undefended (Koch 2003a). Further enhancements to information operations is likely to focus on further integration of information operations with military operations and the development of platforms that provide integrated Electronic Support Measures in addition to Electronic Counter Measure capabilities with previously unachievable location accuracy (*Jane's Defense Industry* 2003).

Potential and challenges of effects based operations

EBO holds promise for the future of warfare as successful execution can allow militaries to economize on the employment of force and reduce the numbers of troops needed on the ground during the high-intensity phase of war. Economy of ground force employment will limit casualties on both sides of the conflict. Similarly, collateral damage in terms of civilian casualties and infrastructure damage can be minimized. However, achieving economy of effort on the battlefield is not a simple task. To do this a whole array of resources are required, starting with the need to have a comprehensive awareness of the battlefield by employing pervasive and persistent sensors; then the ability to manage the knowledge created of both expected enemy courses of action and one's own responses to achieve the desired strategic outcomes; and, finally, the ability to effect those outcomes through the precise application of kinetic and non-kinetic means on the targets of choice.

The range and depth of assets employed by the US during Operation Iraqi Freedom suggest that the acquisition of resources to conduct EBO is costly. The US had attained its current superior military position by outspending everyone else; in fact its 2002 military expenditure is greater than the combined total of the next top 14 spending countries in the world (Sköns *et al.* Yearbook 2004: 312). Inherent in the US philosophy is the choice to spend money in order to save lives on both sides of the battlefield and reduce infrastructure damage. Even then, the US has not been able to fulfill all aspects of EBO due to the weaknesses inherent in existing combat systems. The US has reached a high level of attainment in effects based planning and targeting and possibly the conduct of rapid dominance, but has some way to go in employing all sources of national power in conflict resolution, as evidenced by the post-war difficulties in Iraq.

The quick pace of the war forced the coalition to deal with the post-war situation as soon as they moved into cities and areas once under the dominion of the Republican Guard and Saddam loyalists. There was no clear delineation between the war and the post-war phase as some units of the coalition were already encountering post-war issues whilst other units were still moving to Baghdad. Inadequate planning with respect to the post-war effort has cost the US dearly in terms of lives and money. Indeed, the ability of armies to conduct Krulak's three-block war of conventional conflict, unconventional operations and post-war action, right down to the lowest levels of the hierarchy will be the hallmark of future successful operations (Krulak 1999: 3).

Although some form of collaboration exists through chat rooms, messaging, and Operational Net Assessment (ONA), it is unlikely that collaboration will be fully achieved until all the forces are networked in a common architecture with the ability to track everyone's position automatically. Finally, the US has some way to go in conducting EBO by treating the adversary as a complex adaptive system. The challenge will be to develop of a cultural–military–economic model that is so comprehensive as to be able to foretell what the likely behavior or actions of the adversary will be in response to our own actions.

If the US has some way to go in the conduct of EBO, the other major coalition partners in the operation, namely the UK and Australia are not even in the same ballpark when it comes to the conduct of EBO. These two countries are still very much focused on effects based planning with a limited ability in effects based targeting.

Conclusion

So far, the US approach to EBO has focused mainly on the instrumental, or the technological aspects, but not the existential aspects of war. If we believe that despite our best efforts to instrumentalize war, Clausewitzian fog is inevitable, it would also be necessary then to focus on the human aspects of war, or what will enable the soldier to better operate in a complex environment. Of primary importance is professional military education. The ability of the soldiers to find workaround solutions to problems encountered on the battlefield and to continue to operate despite imperfect equipment was a strength highlighted time and again in lessons on the conduct of Iraqi Freedom.

Hence, the successful conduct of EBO requires capabilities as diverse as PGMs, persistent sensors, and computer models. It also requires one to have capabilities to conduct information operations, special operations and maneuver warfare. If one does not possess the full spectrum of such capabilities, the conduct of EBO is likely to be limited. More importantly, it requires the humans in the loop to know the adversary and self so well as to allow one to effectively dictate the pace of war and render the adversary struggling to keep up. This brings to mind Krulak's idea of the strategic corporal having to exercise an exceptional degree of independence, maturity, restraint and judgment in the conduct of operations in the twenty-first century. So important is this requirement for the mastery of knowledge that some people have dubbed effects based operations "PhD level warfare" (Mills 2003).

References

Alberts, David S., Garstka, John J., and Stein, Frederick P. (1999) *Network Centric Warfare: Developing and Leveraging Information Superiority*, Washington, DC: DoD CCRP Publication Series.

Ames, Ben (2003) "Air Force Predicts Enemy Moves with New Software," *Military & Aerospace Electronics*, 1 September.

Army Magazine (2003) "Operations Iraqi Freedom: A Chronology," May.

Beck, Sara and Downing, Malcolm (eds) (2003) *The Battle for Iraq: BBC news Correspondents on the War against Saddam and a New World Agenda*, London: BBC Worldwide Limited.

Burger, Kim (2003) "What Went Right?" *Jane's Defense Weekly*, 30 April.

—— (2003) "US Army Test Precision-Strike Unmanned Air Vehicle," *Jane's Defense Weekly*, 3 September.

Cerasini, Marc (2003) *The Future of War: The Face of 21st Century Warfare*, United States: Alpha.

Clausewitz, Carl von [Michael Howard and Peter Paret (eds and trans.)] (1976) *On War*, Princeton: Princeton University Press.

Cook, Nick (2003) "Effects-Based Air Operations – Cause and Effect," *Jane's Defense Weekly*, 18 June.

Cordesman, Anthony (2003) "The Lessons of the Iraq War," *Summary Briefing for the Land Warfare Studies Center – Chief of Army Conference 2003*, 2 October.

Crowder, Gary L. (2003) "Effects-Based Operations: The Impact of Precision Strike Weapons on Air Warfare Doctrines," *Military Technology*, 27(6): 16–25.

Deptula, David A. (2001) *Effects-Based Operations: Change in the Nature of Warfare*, Virginia: Defense and Airpower Series, Aerospace Foundation.

Douhet, Giulio (Dino Ferrari, transl.) (1942) *The Command of the Air*, New York: Coward–McCann.

Fayette, Daniel F. (2001) "Effects Based Operations: Application of New Concepts, Tactics, and Software Tools Support the Air Force Vision for Effects Based Operations," *Air Force Research Laboratory's Horizons*.

Filkins, D. and Jehl, D. (2003) "US Moved to Undermine Iraqi Military Before War," *New York Times*, 10 August.

Fulghum, David A. (2003) "Offensive Gathers Speed," *Aviation Week & Space Technology*, 158(12): 22–4.

Gellmann, Murray (1994) *The Quark and the Jaguar: Adventures in the Simple and the Complex*, New York: W.H. Freeman & Company.

Gleeson, Dennis J. (2001) *New Perspectives on Effects-Based Operations: Annotated Briefing*, Alexandria, Virginia: Institute for Defense Analyses.

International Defense Review (2003) "USN Eyes Tactical Tomahawk as 'Call-for-Fires' Weapon," 1 September.

International Institute for Strategic Studies (2002) *The Military Balance: 2002–2003*, London: Oxford University Press.

Jane's Defense Industry (2003) "Lockheed Martin to Lead UK's Soothsayer EW Programme," 1 October.

Kenyon, Henry S. (2003) "Unconventional Information Operations to Shorten Wars," *Signal*, 57(12): 33–40.

Klein, Gary (1998) *Sources of Power: How People Make Decisions*, Massachusetts: MIT Press.

Knights, Michael (2003) "USA Learns Lessons in Time-Critical Targeting," *Jane's Intelligence Review*, 1 July.

Koch, Andrew (2003a) "Information War Played Major Role in Iraq," *Jane's Defense Weekly*, 23 July.

—— (2003b) "Information Warfare Tools Rolled Out in Iraq," *Jane's Defense Weekly*, 6 August.

Krulak, Charles C. (1999) "The Strategic Corporal: Leadership in the Three Block War," *Marines Magazine*, 28(1): 28–34.

Lambeth, Benjamin S. (2002) "Lessons from the War in Kosovo," *Joint Forces Quarterly*, Spring 2002, Issue 30: 12–19.

Liddell Hart, Basil H. (1972) *Strategy*, New York: Praeger.

Mann, Edward, Endersby, Gary, and Searle, Tom (2001) "Dominant Effects: Effects-Based Joint Operations," *Aerospace Power Journal*, 3(15): 92–100.

Mills, Greg (2003) "New War, Fresh Tactics ... and Old Lessons," *The Straits Times*, 27 March.

Morley, Jefferson (2003) "The Origins of 'Shock and Awe,'" *Washington Post*, 21 March.

Noonan, Michael P. (2003) "The Military Lessons of Operation Iraqi Freedom," *FPRI E-Notes*, Foreign Policy Research Institute, 1 May 2003.

PR Newswire (2003) "Raytheon and Lockheed Martin Expand Roles of BAe Systems and General Dynamics in Pursuit of Distributed Common Ground System 10.2 Program," 17 July.

Shanahan, John N.T. (2001) "Shock-Based Operations: New Wine in an Old Jar," *Air & Space Power Chronicles Online Journal*, 15 October 2001. Online, available at www.airpower.maxwell.af.mil/airchronicles/cc/shanahan.html

Sirak, Michael (2003a) "Boeing Wins Small Diameter Bomb Deal," *Jane's Defense Weekly*, 3 September.

—— (2003b) "US Air Force Plans Substantial Increase in Cruise Missile Buy," *Jane's Defense Weekly*, 17 September.

—— (2003c) "USAF Seeks to Combine Laser, GPS in Bomb," *Jane's Defense Weekly*, 8 October.

Sköns, Elisabeth, Perdomo, Catalina, Perlo-Freeeman, Sam and Stålenheim Petter, (2004) "Military Expenditure," *SIPRI Yearbook 2004*, Oxford: Oxford University Press.

Smith, Edward A. Jr. (2002) *Effects Based Operations: Applying Network Centric Warfare in Peace, Crisis, and War*, Washington, DC: DoD CCRP Publication Series.

Strategic Comments (2003) "Lessons from the Iraq War," 9(3): 1–2.

Streetly, Martin (2003) "Airborne Surveillance Assets Hit the Spot in Iraq," *Jane's Intelligence Review*, 1 July.

Sun Tzu [Tao Hanzhang, transl.] (1987) *The Art of War*, New York: Sterling Publishing.

Ullman, Harlan K. (2003) "Shock and Awe Revisited," *RUSI Journal*, 148(3): 10–14.

Ullman, Harlan K. and Wade, James P. (1998) *Rapid Dominance – A Force for All Seasons*, London: Royal United Services Institute for Defense Studies, RUSI Whitehall Paper Series.

Vick, Alan, Moore, Richard M., Pirnie, Bruce R., and Stillion, John (2001) *Meeting the Challenges of Elusive Ground Targets*, Santa Monica: RAND.

Williamson, John (2003) "Bigger, Better C4ISR Systems Underpin US Warfighting Efforts," *Jane's International Defense Review*, 1 August.

5 The transformation of special operations forces in contemporary conflict

Malcolm Brailey[1]

Since the collapse of the Soviet Union in August 1991, armed conflict and the use of military force have transcended many traditional and well-established boundaries. Numerous scholars have grappled with the systemic and strategic nature of these changes. However, significantly less attention has been given to the equally important changes and transformations occurring among the many and varied actors who actually participate in war. This chapter addresses this lacuna, in part, by examining the steadily increasing importance of Special Operations Forces (SOF).

By the late twentieth century, it became almost universal for armed forces to include elite combat units within their organizational structure. Generally, the *raison d'etre* for any of these elite units was to support the aims of conventional strategy and to supplement the activities of conventional military forces. Over the past decade, however, SOF have gradually developed into a potent and indispensable component of modern armed forces outside, and separate to, conventional structures and doctrine. They are becoming increasingly "joint" in nature, as well as displaying great utility across the spectrum of conflict. More often than not, SOF now directly shape the strategy and conduct of military operations in both character and intent.

Historically, "unorthodox" strategy is certainly not a new phenomenon: Guerrilla warfare has long been the tactic of the "irregular" combating the "regular." In this context, however, "irregular" has tended to be associated with non-state or quasi-state groups; and "regular" with the standing and professional armed forces of governments and legitimate political actors. What is most striking about special operations in the late twentieth century is the "extraordinary growth in the irregular activities of the regulars ... to secure strategic effect through an unconventional style" (Gray 1999: 286).

This chapter aims to document and explore this phenomenon in detail. It begins by locating and defining special operations and SOF in the rubric of strategic theory. Then, from a more empirical perspective, the major changes and emerging trends in the missions and capabilities of SOF that have become evident in the many conflicts of the last decade are identified. Finally, a similar empirical framework is employed in order to examine the innovative tactics of SOF evident in more recent operations, such as those in Afghanistan and Iraq, and the equally innovative and adaptive organizational changes. The argument is that SOF represent a harbinger for change in the way states think about the conduct of war. In addition, they may offer an increasingly viable, effective and legitimate alternative to traditional approaches to the use of force by states. Most of the argument is drawn from an analysis of the transformation and deployment of SOF from two developed nation-states at either end of the defense material scale: the United States (US) and Australia. These two nations have greatly advanced special operations in recent years. Thus, they provide a relevant and salient example of how these forces have been transformed and of future possibilities.

Strategic theory and SOF

It is necessary to preface the wider examination of SOF with a brief foray into what this term means from a theoretical perspective. Generally, the study of these forces suffers from being an inherently rich empirical subject with little or no available theory of contemporary strategic relevance. A rudimentary look at the strategic fundamentals and principles of both "SOF" and "special operations" will provide added nuance to the empirical information presented in this chapter.

SOF tend to be the part of any defense force that is glorified by the media and the entertainment industry, but remain shielded behind the closed doors of government security policy and, more often, self-imposed secrecy. Also, they generally consist of unique elements specific to particular national defense requirements. Therefore, they exhibit few commonalities internationally other than a shared elite status. As such, the initial temptation in defining SOF is to juxtapose their identity against the mainstream or conventional military identity – the special – as opposed to the general-purpose forces that make up any defense force. This tautological approach is, however, largely inadequate for the purposes of a wider study of these forces from a strategic theory perspective. It would include too wide a variety of military organizations with very different missions and capabilities. For example, some special units may fulfill internal policing and intelligence roles, while others may focus on a particular specialization or military capability such as parachuting or mountain warfare. This situation highlights national differences rather than international commonalities and reveals little about their higher strategic function. An alternative and more fruitful approach is to first define what constitutes a "special operation" and to extrapolate from that which kinds of forces are selected and trained to perform those operations on a case-by-case basis.

Defining special operations

This chapter defines SOF as those discrete elements of legitimate state-based military forces that possess a unique military capability to conduct "special operations" that require personnel who are specifically selected, trained and organized to conduct these special operations.

In Western military thought, special operations are most commonly defined in the context of conventional high-intensity war, as shaped by the experience of significant interstate conflict during the twentieth century. For example, Luttwak (1982: 1-1) describes special operations as "self-contained acts of war mounted by self-sufficient forces operating within hostile territory." Likewise, Foot (1970: 19) sees special operations as "unorthodox coups ... unexpected strokes of violence, usually mounted and executed outside the military establishment of the day, which exercise a startling effect on the enemy: preferably at the highest level."

However, Tugwell and Charters (1984: 34–35) have correctly noted that these definitions are deficient because they exclude special operations that are undertaken outside the context of conventional war; that is, without a well-defined enemy, frequently not in hostile territory (though arguably still danger-ous), and indeed they may not always involve the use of violence. Rather, the authors consider special operations to be "[s]mall scale, clandestine, covert or overt operations of an unorthodox and frequently high-risk nature, undertaken to achieve significant political or military objectives in support of foreign policy" – a definition which still remains salient today. Further, special operations are characterized by "either simplicity or complexity, by subtlety and imagination, by the discriminate use of violence, and by oversight at the highest level. Mili-tary and non-military resources, including intelligence assets, may be used in concert."

The US Department of Defense (DoD) has more recently affirmed the scope of this earlier academic position with a similar policy definition of its own. It states that:

> [s]pecial operations are operations conducted in hostile, denied, or politic-ally sensitive environments to achieve military, diplomatic, informational, and/or economic objectives employing military capabilities for which there is no broad conventional force requirement.
>
> (Department of Defense 2004)

The strategic emphasis of special operations is also evident in recent Australian military doctrine. Australia defines special operations as highly focused operations executed at the tactical level, using unconventional military means, designed to achieve wider operational and strategic effects. Importantly, Australia also acknowledges that special operations are shaped by political and military consider-ations, and that they therefore require oversight at the national level.

At the tactical level SOF obviously can have a direct influence, at the opera-tional level these forces can have both a direct and indirect influence; and at the strategic level they can also have an indirect influence. Gray (1999: 168–185) has argued that the use of special operations and the recent concurrent expansion of SOF may be attributed largely to a two-fold strategic utility. First, these forces provide economy of force – a key principle of war – because they can achieve significant results with limited forces. They act as a force multiplier on the battlefield for other conventional components and can, relative to their size, also have a disproportionate impact on a battle themselves. Second, SOF can expand the options available to political and military leaders in support of their respective goals. While in theory, there are always alternatives to the use of force – for example, diplomacy and sanctions – in practice there will always be situations where force or physical coercion may be unavoidable. SOF give states a flexible, precise and minimal capability to impose their will on their opponents. In the context of today's increasingly complex security environment, Mark Mitchell (1999: 84) has pointed out a further, vital claim for the strategic utility of SOF; "tailor-to-task capabilities." Indeed, today's SOF seem able, given their extant skills and operational maturity, to adapt to a wide variety of constantly changing situations and conditions.

Missions and capabilities

It would be incorrect to imply that contemporary SOF are fundamentally different from either their historical antecedents or recent forerunners. Irregular units or raiding forces have been evident in the methods of warfare of many societies for millennia (Moreman 1992: 35–64). Nevertheless, as Gray (1996: 146) noted, the "systemic organization and training of small elite groups of soldiers ... is essentially a recent innovation in warfare." In this modern context, SOF undoubtedly experienced their greatest expansion and first rise to prominence in World War II. In that conflict, almost all involved parties developed and employed some kind of unorthodox or irregular forces. Since then, and this appears to be a trend of increasing importance, SOF have been called upon to conduct an ever-widening array of missions that are demanding the acquisition of new skill sets. While specific roles are generally country-dependent, several key areas can be highlighted as significant in the recent wider development of SOF missions and capabilities. They include: offensive counter-terrorism, domestic counter-terrorism, non-traditional operational roles, and unconventional warfare operations.

Offensive counter-terrorism

SOF have most recently been required to conduct globally dispersed pre-emptive military operations, especially in the context of post September 11 counter-terrorism missions. Indeed, SOF have become the force of choice for this operational challenge. The utility of SOF for this mission lies not only in their warfighting, capabilities, but more importantly in their high level of non-warfighting capabilities, especially their language skills, politico-cultural specialties, and their mastery of information and communications technologies. A recent RAND report has noted that this offensive orientation is markedly different from past counter-terrorism efforts, such that this orientation can be defined as offensive counter-terrorism to better distinguish it from more traditional doctrinal tasks (Nardulli 2003: viii).

It would seem that only the US military has conducted offensive counter-terrorism missions on a global scale since September 11, although it is likely that other coalition and partner nations have participated and cooperated with the US on such tasks. The exact nature and number of offensive counter-terrorism missions remains unknown, but there is substantial secondary evidence to suggest a high tempo of activity in the Global War On Terror and in the hunt for weapons of mass destruction. For example, William O'Connell, the US Assistant Secretary of Defense for Special Operations and Low Intensity Conflict, has pointed out that US SOF are currently conducting "combat missions, strategic reconnaissance ... and training operations worldwide" (Gilmore 2003). Feickert (2006) has also recently noted that US SOF continue to operate in Iraq and Afghanistan where they are "actively pursuing key insurgents." In addition to these high-profile activities in Iraq and Afghanistan, US SOF and partner nations have also been concurrently involved in offensive counter-terrorism operations in the Philippines, Djibouti, Yemen, Somalia, Pakistan, Georgia, Uzbekistan and Columbia (Corera 2003). In terms of operational success, since September 11, more than 3,000 operatives have reportedly been captured in more than 100 countries. In the same period, more than 50 terrorist leaders and planners have been either killed or captured in 20 different countries (Billingslea 2003).

Perhaps not surprisingly, the former US Secretary of Defense, Donald Rumsfeld, had formally given responsibility for offensive counter-terrorism operations to the US Special Operations Command (Fitzsimmons 2003: 203–218). Secretary Rumsfeld subsequently also designated US Special Operations Command (USSOCOM) as the "global synchronizer" in the war on terrorism for all the US military combatant commands, and gave them responsibility for designating a new global counter-terrorism campaign plan and conducting preparatory reconnaissance missions against terrorist organizations around the world (Robinson 2005). This new approach, documented in the classified National Military Strategic Plan for the War on Terrorism, unified the US military under a special operations umbrella for the first time. This is a significant development, since it is now a dedicated Special Operations Force headquarters that has primary control over all offensive counter-terrorism operations worldwide.

Domestic counter-terrorism

Counter-terrorism also has its missions and roles in the more traditional domestic environment. Many nations have maintained – and in some cases very successfully used – domestic counter-terrorism capabilities for some time. However, typically, the military has been seen as a force of last resort, with primary responsibility instead resting with law-enforcement agencies or paramilitary forces. Once again, however, September 11 caused a paradigm shift in this regard. Since then, extremist and religiously motivated terrorism has increased in scale, frequency and lethality. This means that the standard law-enforcement response may now be largely inadequate. The much higher level of operational capability found in SOF now seems to be the more natural, and logical, response to domestic terrorist incidents.

The demand for "homeland security" has indeed evolved into a new strategic milieu, which has seen governments and their citizens adjust their perceptions about the boundaries of national defense and security policies. In this context, military and SOF involvement has come to be seen as a defining part of any homeland security strategy. Australia, for example, has effectively doubled its domestic counter-terrorism capability since September 11. In May 2002, the Australian Defense Minister, Robert Hill, announced the details of several new counter-terrorism initiatives. The most significant of these initiatives was the raising of a second Tactical Assault Group (TAG) at a cost of A$219.4 million (Minister for Defense 2002). The second TAG is based in Sydney and comple-ments the existing group located in Perth (Lewis 2003a: 45–52). Both groups are primarily designed to resolve hostage, or siege-style, terrorist incidents beyond the organic capabilities of Australian state police. This task is broadly known as "special recovery." TAG capabilities also extend to a wider range of tasks including service-assisted or protected evacuation, entry from the air and sea, maritime point of entry, and combat search and rescue (Minister for Defense 2002). Jones (2002) believes it is almost inevitable that any major act of terror-ism occurring within Australian territory, or against Australian interests, would elicit an Australian Defense Force response. The doubling of the Tactical Assault Group capability is a further important signal in that regard.

The Australian SOF community is now firmly a part of the Australian Government's homeland security strategy. It provides policy advice both inside and outside the Department of Defense, contributes to national counter-terrorism command and control arrangements, and maintains an increased operational capability to combat terrorism within the Australian domestic environment (Lewis 2003a: 45–52). The Australian Government clearly recognizes the growing importance of SOF in Australia's security policy arrangements. This recognition was highlighted when in 2004 it appointed Duncan Lewis, the former Special Operations Commander Australia, as the First Assistant Secret-ary for National Security in the Department of the Prime Minister and Cabinet (Minister for Defense Media Release 2004). This important civilian appointment plays a key role in driving homeland security and counter-terrorism policy in Australia, and the appointment of a former senior special operations officer is a significant development in Australia's policy approach.

Similar developments in the US are somewhat more difficult to detect given that domestic counter-terrorism has long been the responsibility of the Federal Bureau of Investigation's Hostage Rescue Team (Guttieri 2003). However, the recently established US Northern Command has been assigned to deter, prevent and defeat threats and aggression targeted at the US and its territories. This mission includes Operation Noble Eagle, an ongoing, US-based homeland defense and civil support operation associated with the wider war on terrorism (Bolkom 2003: 3). It can be reasonably assumed that Northern Command will at some stage have access to the SOF normally reserved for offshore incidents in situations requiring a military response to domestic terrorist incidents. The US Special Operations Command (USSOCOM) has also recently established the Counter-Terrorism Campaign Support Group, whose mission is to provide inter-agency, civil, and military support to existing federal agencies at the operational level. For USSOCOM, domestic counter-terrorism has become its second operational priority, after "pre-empting global terrorism" and "chemical, biological, radiological, nuclear and explosive threats" (Department of Defense 2004: 29).

Non-traditional missions

Since the mid-1990s, SOF have also been tasked with a wide variety of non-traditional missions outside the warfighting rubric. Initially, it is likely that governments were drawn into using, or choosing, these forces due to the latter's high levels of operational readiness, agility, and broad range of capabilities as opposed to conventional forces. Over time, the generally outstanding performance of SOF on non-traditional missions has made them increasingly the tool of choice for policy makers. Since the 1990s Australia and the US have used SOF extensively for non-traditional missions. Nowhere is this more obvious than in counter-narcotic operations and the capture of terrorists and war criminals.

In recent years the US Government has deployed its SOF in covert and overt missions in South and Central America, in support of broader national and international anti-drug policies (Department of Defense 2004: 40). This is especially useful in combating drug cartels whose armed elements have evolved from classic guerrilla groups into terrorist and drug-trafficking organizations, leading some scholars (for instance Valenzuela and Rosello 2004) to call for military support commitments based on doctrinal Army Mobile Training Teams that were used very successfully in El Salvador during the 1980s and early 1990s. In 1997 alone, US SOF conducted some 194 anti-drug missions, mostly in Central America (Schoomaker 1998: 5). These forces remain in the region, and are "continuously training" host-nation counter-narcotics forces, particularly in Ecuador and Colombia, where Colombian military forces recently captured the key rebel leader and drug baron Ricardo Palmera. He was allegedly captured with the assistance of US SOF (Forero 2004).

Australian SOF have also begun to play a key role in major anti-drug operations. On 20 April 2003, Australian special operations force elements boarded the *Pong Su*, a North Korean ship in Australian territorial waters suspected of drug trafficking. The boarding team included members of the Special Air Service Regiment, Army commandos and Navy clearance divers. The ship was subsequently found to have smuggled 50 kilograms of heroin into Australia, and the operation was a major success for the Australian Defense Force (ADF) and the various law enforcement agencies with which they were cooperating (Conford and Malkin 2003). It is significant that only special operations force assets were able to carry out such an assault, and at extremely short notice, with little or no preparation. The Special Air Service Regiment, in particular, has long been developing an offshore recovery capability for domestic counter-terrorism missions. This capability is now of increasing appeal to state and federal law-enforcement agencies planning anti-drug operations.

In the history of SOF, locating and capturing important political and military leaders has been a major task. An example is the capture of Ah Hoi, the Malayan Chinese communist leader, by the British Special Air Service in the jungles of Malaya in 1958 (Conner 2000: 50). More recently, many SOF have been tasked with missions against war criminals and terrorist leaders. These missions have several unique characteristics. First, the task of finding and capturing these new targets typically does not take place in a well-defined area of operations. Second, the requirement for intelligence leading to a successful mission is also typically interagency or international in nature, requiring high levels of coordination and cooperation. Third, SOF are most likely to be operating outside the normal legal conventions governing international armed conflict, despite the often public and political rhetoric to the contrary. This environment affects both their legal status and that of their targets.

This role has been particularly notable in Bosnia-Herzegovina, where many International Criminal Tribunal-indicted war criminals from the former Yugoslavia remained at large following the signing of the Dayton Peace Accords in 1995. The task of finding and arresting them was left largely to US and UK SOF. Typically, these forces have operated outside the NATO mandate, conducting national direct-action or special-recovery tasks. US Special Mission units – such as SEAL Team 6 and Delta Force – have operated sporadically in Bosnia since 1997 (Newman 1998). The UK Special Air Service Regiment has maintained a small permanent detachment in Bosnia searching for war criminals. This permanent presence has given its members the distinct advantage of being able to gather their own intelligence over time, plan and train for specific missions, and also to react rapidly on any local information; as was the case when the Special Air Service captured General Stanislav Galic, the Serb commander who besieged Sarajevo, during December 1999.

US SOF also maintain a substantial strategic-reconnaissance/special-recovery capability. In Somalia between 1992 and 1995, US Special Mission units have been successful in apprehending key members of Mohamed Aideed's leadership infrastructure (Celeski 2002: 16–27). More recently, the 2003 war against Iraq saw a much larger and more concerted effort to use SOF in the search for the Baathist leadership. A new "covert" force called Task Force 121, apparently a unique and tailored special operations task unit that has been designed to act with greater speed on intelligence tips about "high-value targets," has allegedly been created to "hunt down" former Iraqi leaders and key terrorist operatives across the region (Hersh 2003; Shanker and Schmitt 2003). US Special Mission units may have also returned to southeastern Afghanistan, along the Pakistan border (Gellman and Linzer 2004), as part of a new effort to find Osama bin Laden and his remaining leadership in Afghanistan, and to counter a resurgence of Taliban and al-Qa'ida forces there. Strong evidence as to the ongoing nature of these missions in Afghanistan emerged in the tragic events surrounding the crash of a US special operations Chinook helicopter in the Kunar province in

2005, and the subsequent loss of all 16 troops on board. Those special operations troops were engaged in a rescue mission for a smaller group of four US Navy SEALS, who were evidently involved in an operation to capture a key Taliban commander with close links to al-Qa'ida.

Unconventional warfare operations

Finally, one of the most interesting and challenging missions that SOF are increasingly assigned to perform is that of unconventional warfare, which constitutes a "broad spectrum of military and paramilitary operations ... conducted by, with, or through indigenous or surrogate forces who are organized, trained, equipped, supported and directed by an external source" (Dickson 2001: 16–17). Ironically, unconventional warfare was one of the original missions that led to the establishment of standing special operations units in modern Western armed forces. This particular capability developed out of the extensive experience that the Allied nations gained during World War II in training and equipping partisan and guerrilla forces ranging from France and Yugoslavia in Europe to Burma and Timor in Asia (Jones and Tone 1999: 5–6).

Since the early years of the Vietnam War, however, most Western special operations units have only rarely practiced true unconventional warfare. Certainly by 1965 it had largely become a low-priority "legacy mission" (Erckenbrack 2002: 8). In the early twenty-first century, unconventional warfare is once again a focal point for special operations and military capabilities. This renewed focus is driven largely by the spectacular success of predominately US SOF in Afghanistan during late 2001 and early 2002 on Operation Enduring Freedom. Here, the attraction of using SOF in an unconventional warfare role was twofold. First, in the words of the former Secretary of Defense Donald Rumsfeld, "you don't fight terrorists with conventional capabilities, you do it with unconventional capabilities" – and the specialized combat skills that only special operations personnel could provide (Kennedy 2002). Second, the US political and military leadership wanted to avoid repeating the past bitter experiences that resulted when large numbers of British and Soviet conventional ground troops had been deployed into Afghanistan in the past.

The most famous instance of the success of SOF in unconventional warfare was undoubtedly the liberation of the city of Mazar-e Sharif on 10 November 2001. During the liberation, members of the US 5th Special Forces Group helped the Northern Alliance defeat vastly superior Taliban forces (Biddle 2002: 8–10). This pattern of success continued, with advisers from the US helping the Northern Alliance in the liberation of almost every major city in Afghanistan, including Kabul, Jalalabad, Konduz and Kandahar (Kennedy 2002). The 5th Special Forces Group had been conducting a similar unconventional warfare mission in another Central Asian country for about six weeks before the attacks of 11 September 2001. Its members were rapidly redeployed by mid-October when they subsequently linked up with Harmed Karzai and his Northern Alliance forces (Finn 2001). The members of this special operations unit all had significant operational experience in Central Asia, developed over many years of deployments, and many spoke local or regional languages. These attributes demonstrate the long-term approach and commitment required to develop the range of unconventional warfare skills and capabilities among special operations personnel.

In recent years, unconventional warfare has undergone a revival in terms of its place in US defense policy. In that context, SOF are now seen as "Global Scouts" who serve to "assure US allies and friends of US government resolve" (Department of Defense 2004), and who in the future will be used to defeat improved enemy "means and methods of anti-access and anti-denial" activities (Erckenbrack 2002: 8). Even before recent unconventional warfare missions conducted in support of the Global War On Terrorism, the US was widely using its SOF in training and assistance missions worldwide. In 1997 alone, US SOF were deployed to 144 countries (Schoomaker 1998).

In Afghanistan, the success of special operations missions may lead to the "SOF-centric" campaign being regarded as a possible future model of warfare, applicable across a wide range of future conflict types. In Operation Iraqi Freedom, the Combined Forces Special Operations Component Command coordinated the activities of Kurdish irregulars in northern Iraq as a key aspect of the overall operational strategy (Fitzsimmons 2003: 207). Elsewhere, the Australian Special Air Service Regiment reportedly developed an "UW Wing" in the early 1970s, modeled on the US Army "A" Team concept, and initially using US doctrine. Horner (2002: 398–404) has documented the Special Air Service Regiment's involvement in pseudo-unconventional warfare missions throughout the 1970s and 1980s. Examples of these missions include Special Air Service Regiment personnel training army and security elements within Thailand and Indonesia.

The unconventional warfare mission is currently overshadowed by the Global War On Terror and offensive counter-terrorism operations. However, it may provide an extremely effective method of dealing with state-based threats, weak states, and even non-state actors into the future. Newman (2005: 28–36) believes that, in the war on terrorism, unconventional warfare conducted by SOF attacking non-state actors is an extremely attractive alternative to traditional notions of applying military force. Newman highlights reduced resource requirements, economy-of-force advantages, manageable moral obligations and demonstrated historical successes among the reasons why an unconventional warfare approach can effectively interdict global targets without the need for massive conventional build-ups.

Organization and tactics

The second major element contributing to the transformation of SOF is how these forces are organized, trained, and how they operate. The nature of special operations, at least since World War II, has generally demanded a combination of land, air and sea assets operating simultaneously in a multidimensional fashion. As such, while most conventional armed forces cling to their single-service traditions, many SOF have transformed themselves into truly joint-force organizations. This new emphasis means that, in terms of command and control (and even tactical cooperation at times), SOF now reflect capability rather than service lines. As a result, these forces are now at the forefront of the many truly coalition operations and combined missions conducted as part of the global war on terror and international peacekeeping.

Joint organizations

The theory and practice of joint warfighting refers to the "synergistic application of the unique capabilities of each service so that the net result is a capability that is greater than the sum of its parts" (Noonan and Lewis 2003: 31). Two factors drive the desire for joint operations: the natural advantages of military efficiency; and changes in the global strategic context, which stem from a blurring between the levels of war and military operations other than war, and are driven by a civil–political requirement for more precise applications of combat power globally (Noonan and Lewis 2003: 33). Joint warfighting presents a series of interoperability problems in four key areas: culture, technology, compartmentalization and organizational structures (Call 2003: 8).

First, special operations are usually multidimensional by nature and therefore demand the involvement and cooperation of land, maritime and air power elements. SOF therefore become increasingly interoperable with conventional forces in the true sense of the word (Call 2003: 7). This interoperability has been achieved over time through the conduct of regular joint training and operations, both within the special operations community and with conventional forces. SOF therefore have a strong operational legacy of planning and executing joint missions across a wide spectrum of conflicts. Furthermore, they also have a strong legacy of inter-agency cooperation with domestic border control and law enforcement agencies, as well as international organizations such as the UN, non-governmental organizations and even private military companies. Internationally, they have demonstrated their understanding of joint warfare concepts and doctrines and their ability to practice these in training, and on operations, by implementing genuine, joint organizational structures.

The US special operations community has been at the forefront of joint doctrine, training and organization for well over a decade. In 1986 the US Congress expressed concern for the status of SOF within overall US defense planning. This concern arose largely as a result of shortcomings identified in the failed Iranian hostage rescue attempt in 1979, Operation Eagle Claw; and compatibility and unification problems from the Grenada invasion in 1983, Operation Urgent Fury. These concerns led directly to the creation of USSOCOM, authorized by the Cohen–Nunn Amendment to the Department of Defense Authorization Act of 1987 (Schoomaker 1998: 3). This law mandated the creation of a unified command with "service-like" responsibilities to oversee all SOF and report directly to the Secretary of Defense for all budget, equipment, training and doctrinal issues.

USSOCOM is now one of the nine unified combatant commanders. As such, it is responsible for planning, directing and executing special operations and providing special operations units to support the other geographic combatant commander's theater security cooperation plans (Feickert 2003: 3). There are approximately 49,000 active and reserve personnel in USSOCOM. They are sub-organized into three component commands, which comprise the major units and training establishments, namely the United States Army SOF, the Naval Special Warfare Command and the Air Force Special Operations Command. On 1 November 2005 DoD announced the creation of the Marine Special Operations Command as a new component of USSOCOM (Feickert 2006: 3). One sub-unified command also exists – Joint Special Operations Command (JSOC) – which provides a joint headquarters to study special operations requirements, ensure compatibility between services and equipment standardization, develop joint doctrine and tactics, and conduct joint exercises and training (Department of Defense 2004: 12–13). JSOC is also widely believed to command and control what are described as the US military's three "special mission units" (Feickert 2006: 4). Since 1990, the US special operations community has also implemented, at the lower tactical and operational levels, innovative solutions to the requirement for liaison and coordination with other conventional force elements. This approach has enhanced compatibility and synchronization of all force elements (Call 2003: 11–16).

Since 2003, the ADF has transformed the way in which Australia's SOF are organized. The Australian Government created a new Special Operations Command (SOCOM) on 5 May 2003, partly as a response to the October 2002 Bali bombings (Minister for Defense 2003). The creation of SOCOM confirmed the Government's intention to enhance the Australian Defense Force's special operations capability and to meet the increasing need for an effective joint, inter-agency and alliance, counter-terrorism and anti-terrorist capability (Lewis 2003a). SOCOM is a true joint headquarters, with component command status equivalent to the Maritime, Land, Air and Joint Logistics Commands. SOCOM reports directly to the Chief of the Defense Force for counter-terrorism operations, and to the Chief of Joint Operations for special operations support to all other operations. The command consists largely of land units and supporting elements, such as organic logistic and limited aviation support. It is also working towards better cooperation with relevant Royal Australian Navy capabilities and Air Force support. In addition, SOCOM maintains an element of its headquarters in Canberra, to provide future capability development, strategy, and doctrine development support to the broader defense community; to act as a coordination node for counter-terrorism operations; and to maintain a link to other government bodies and organizations (Lewis 2003a).

Joint warfighting

Perhaps the most striking example of the joint nature of SOF, and their role in wider joint warfighting, may be found in the conduct of Operation Iraqi

Freedom in early 2003. Cordesman (2003) has claimed that between 9,000 and 10,000 US special operations members were specifically deployed into Iraq. This deployment accounted for approximately 8 percent of the total combat forces. Units from all three USSOCOM component commands were present in Iraq – including the previously mentioned special mission units – as well as significant force elements from the Australian and UK special operations communities. These multinational forces were grouped into a Combined Forces Special Operations Component Command, which reported directly to the US Central Command as the geographic combatant command. SOF operating in Iraq displayed and used joint warfighting doctrine and principles on several levels.

First, special operations units themselves are inherently joint organizations. Even multinational coalition partners were fully integrated into a single command structure, albeit with their own national command elements. Second, these units independently conducted specialized and small-scale joint operations – for example, the capture of airfields, the securing of offshore oil terminals, and the aforementioned search for Iraqi leadership targets – what Noonan and Lewis (2003: 37) consider to be a defining characteristic of modern warfare. Third, SOF contributed, often as a key component, to wider joint operations at the theater level in combination with air and land elements; and further through interagency cooperation with non-military organizations. Examples of such contributions include: the now-famous "Scud Hunting" missions conducted by Australian and UK Special Air Services in Western Iraq (Ripley 2003); the creation of unique combat teams comprising SOF with small elements of armor – in one case including main battle tanks; other major land force units, such as the US 173rd Airborne Brigade and the UK 45 Royal Marine Commando (Noonan and Lewis 2003: 31); and missions working closely with, or attached to, the Central Intelligence Agency (Cordesman 2003: 362–363).

Combat operations in Iraq saw a maturing of joint force operations whereby the effective integration of conventional forces with SOF capabilities allowed them to nullify the perceived asymmetric threats while permitting precision targeting within the battlespace.

Indeed, the Combined Forces Special Operations Component Command in Iraq was fighting on three fronts simultaneously, with each mission having different objectives and task-specific requirements. In northern Iraq, special operations units were the supported forces. Their mission was to prevent Iraqi units from reinforcing Baghdad. In western Iraq, they supported the air component mission to prevent SCUD launches. Finally, in the south, SOF also supported coalition land forces in their rapid advance to Baghdad. These missions required new approaches to joint forces integration that may serve as models for the future.

SOF and air power

The ability of SOF to conduct joint operations has also led to concrete tactical innovations. Potentially the most important, and successful, advancement has been the integration of airpower with land forces. This integration has involved the use of tactics, techniques and procedures developed to enhance SOF acting as "human sensors" for air missions. It also involves using air power to provide joint fire support to SOF acting as a ground maneuver force (Findlay *et al.* 2003: 10). In terms of specific tasks, the US Joint Special Operations Doctrine specifies two missions: direct-action and special reconnaissance. In direct-action missions, strategic and operational targets are "illuminated" and then destroyed by air-delivered precision-guided munitions. In special reconnaissance support missions the forces provide not only target acquisition, but also area assessment and post-strike reconnaissance or Bomb Damage Assessment data (Department of Defense 1993: II-1). Hester (2003: 3) believes that the joint-force commander can individually deploy SOF as an alternative to air attacks during the joint targeting process.

The need for this ground/air capability stems from the reality that, while precise air attacks against stationary targets has become routine, locating and prosecuting attacks on mobile and time-critical targets remains a much more difficult task. The need for more human sensors on the ground was further demonstrated during the air campaign against Serbia in the late 1990s. Static targets were easily destroyed, but air attacks to destroy or disrupt mobile or concealed Serbian forces, particularly in Kosovo, were more problematic. The US and the UK reportedly provided covert support, including special operations personnel, to the Kosovar Liberation Army during that campaign. Cordesman (1999: 250–256) attributes the limited success achieved by US air power in Kosovo, of which ground targeting for air strikes was a vital part, to that covert support.

Since Kosovo, the tactics and technology associated with joint SOF and combat air power operations have evolved significantly. This strategy has now been successfully proven in combat during both Operation Enduring Freedom and Operation Iraqi Freedom. Sullivan (2002: iv) has argued that such operations in Afghanistan are "a transformation in the operational art of employing forces." For the US, the origins of this new synergy may be traced to the *Joint Vision 2010* concept of effects based precision engagement. The air war, particularly over Afghanistan, demonstrates that US SOF, and those of coalition partners, have made significant progress in joint operational integration. Sullivan (2002: 9) points out that the change in the nature and conduct of air operations in Afghanistan – and now further demonstrated in Iraq – resulted directly from "advances in technology and evolutions in joint doctrine."

Joint command and control

Tactical command, control and coordination of both SOF and combat air assets, was also innovative. Targeting and fire support in a joint environment are normally highly complex processes, made even more challenging by the non-contiguous nature of the modern battlefield. During Operation Enduring Freedom, special operations teams and headquarter elements could not accurately predict the locations of opposition groups or mobile targets. This situation meant that interdiction missions could not be planned in advance and that there was no clearly defined area of operations in which targets could be prerecorded. Throughout the operation Coalition forces used several tactics to overcome these problems.

First was the use of gridded areas of operation and "kill-boxes." Traditionally, operational design has always included two fundamental components: a mission, and a designated area of operations in which to accomplish that mission. This neat battlespace geometry does not always exist in a non-contiguous battlefield such as that in Afghanistan. A series of fire support coordination measures – for instance, no-fire and restricted-fire areas, and kill boxes – were therefore developed to overcome this lack of symmetry (Jackson 2003). Second, rather than pre-planned fire support, there was an increased use of air interdiction initiated and directed by SOF against enemy targets of opportunity the latter chanced upon. This tactic was used in Operation Iraqi Freedom in the "Scud hunting" air missions (Bostock 2003). The ability to call on combat air support meant that special operations units could operate in remote areas without the need for artillery or other land-based fire-support elements (Ripley 2003).

The capacity of SOF to identify and destroy command and control and Scud missile threats was enhanced by the use of other joint Intelligence Surveillance and Reconnaissance assets – such as unmanned aerial vehicles. These vehicles routinely operated with SOF to find, fix, track and target such threats. Streaming video was then typically sent to other combat air platforms to provide targeting information during the engagement phase (Bradley 2004).

Further, at the individual level, special operations teams used Blue-Force tracking devices extensively, which increased situational awareness and reduced the possibilities of fratricide. Tracking devices allowed special operations teams to be fully included in all friendly battle plans. These devices also allowed the teams to liaise more closely with other ground forces in a particular area, and most importantly, to be identified, and avoided if necessary, by friendly aircraft (Cordesman 2003: 363).

Network-centric warfare and technology

Unmanned aerial vehicles and Blue-Force tracker devices are excellent examples of SOF's increasing reliance on high technology. Individual operators and small teams now have access to unprecedented levels of battlefield communication, shared intelligence, and situational awareness. Network-centric warfare uses technologies and tactics that take full advantage of all available information on the battlefield. It enables the rapid and flexible deployment of all available combat assets. SOF appear to have wholeheartedly embraced network-centric warfare, so much so that the outcome of many special operations is now shaped at all levels by the use of high-technology devices.

There is a common misconception that network-centric warfare is merely the electronic linking of various computer systems. In reality, it is far more than that. It comprises both human and technological factors. A better way to conceptualize network-centric warfare may be to understand it as a "powerful set of warfighting concepts and associated military capabilities" that involves "networking three domains of warfare (the physical, information, and cognitive domains) so as to generate increased combat power by: achieving greater speed of command; [and] increasing lethality, survivability and responsiveness" (Department of Defense 2001: 3–10). Looking beyond even that definition, (Alberts *et al.* 2000: 88) describe it as being about:

> human and organizational behavior ... [network-centric warfare] is characterized by the ability of geographically dispersed forces to create a high level of shared battlespace awareness that can be exploited via self-synchronization and other network centric operations to achieve [the] commanders intent.

SOF epitomize that doctrinal vision and make it an operational reality. The efficiencies of network-centric warfare result from extending the sensing ability of an individual entity to the cumulative ability and reach of the entire network, increasing overall combat power and accelerating decision cycles. Gagnon (2002) has argued that network-centric warfare can improve the probability of special operations mission success in three ways: simplicity and innovation in planning; security and repetition in preparation; and speed, surprise and adaptability in execution. Most of the specific network-centric warfare concepts and capabilities that SOF have developed and deployed on recent operations focus on shared situational awareness, robust communications and better sensor-shooter linkages. This new tactic has become so important that even USSOCOM insiders believe that it has dramatically changed the way that SOF conduct their missions (Ackerman 2003: 17–21). The three operational realities of network-centric warfare – situational awareness, precision fires, and growing transparency in the battlespace (Lewis 2003b: 5–8) have also given SOF increasing prominence in the resolution of international conflicts.

Most special operations teams now regularly deploy on operations with accurate digital maps, real-time information on the disposition of friendly and enemy forces in their area, and connectivity to supporting forces located throughout their own battlespace. These systems directly contribute to a network-centric warfare approach. Several communications and information systems were big winners for US and coalition SOF in Operation Enduring Freedom and Operation Iraqi Freedom (Ackerman 2003). The AN/PRC-148 Multiband Inter/Intra Team Radio provided embedded and secure communications between dispersed members in special operations teams. The Multiband Multimission Radio was also important. Iridium handheld satellite telephones with secure sleeves also proved to be invaluable for diverse special operations elements. They were used by units conducting split operations in rugged terrain and for communicating with other government agencies and local allied troops.

Inmarsat provided satellite connectivity in remote locations. The Deployed Node-Light Terminal provided secure data and voice capability. It also permitted teams to dial into the US Department of Defense Secret Internet Protocol Router Network (Ackerman 2003).

An unexpected requirement in Afghanistan was for videoconferencing. Small, briefcase-sized units were deployed for this purpose. A ruggedized scalable suite of computers, network gear and associated software was also essential to mission planning and situational awareness. The Tactical Local Area Network system formed the hardware base, and the Special Operations Digital Environment software package provided battlefield information, intelligence, collaboration and mission-planning tools. This online approach enabled field communication with systems maintained in the continental US. However, it must be noted that technological innovation has its own hazards, particularly for special operations. An example is the tragic death in Afghanistan of several US special operations personnel in late 2001. In this case, an operator typed in the global-positioning system coordinates of a target into his laptop but had to change the battery before he relayed the information. The battery installation allegedly caused a software glitch and the laptop gave the operator's own position as the target to a circling US fighter, resulting in a tragic instance of fratricide (Cook 2002).

Conclusion and implications

The transformation of SOF over the past decade has occurred incrementally at the strategic, operational and tactical levels. This transformation has been so extensive that special operations now form the centerpiece of strategic planning, operational design and tactical execution in response to many contemporary national security threats encountered by the nation-state. The numerous unorthodox combat operations being conducted in the Global War On Terror – where forces are fighting an amorphous and elusive enemy on a global scale in Afghanistan, Iraq and elsewhere – are all demonstrations of this transformation.

As this chapter has shown, governments have asked SOF to significantly expand their capabilities and range of missions since the mid-1990s; and even more so since 11 September 2001. These forces now perform a range of new missions involving offensive counter-terrorism, homeland security, and even a resurgence of the traditional tasks of unconventional warfare. In the process SOF have further shaped their own future by successfully and innovatively restructuring their organizations, and updating and adapting their doctrine, training and tactics. In order to facilitate these changes, SOF have remained flexible and resourceful, with much of this innovation coming from within. Further, innovation has not been driven by technology alone – though that has clearly been a powerful enabler – but more by the intellect and vision of internal personnel.

The significance of this transformation is manifested on several levels. The conduct of contemporary special operations in places like Afghanistan and Iraq may provide both military planners and politicians with a new model for prosecuting armed force against an opponent, be it state or non-state. SOF maintain key capabilities that allow a strategic outcome to be achieved with fewer troops and resources, a lower profile and more certainty of success. The combination of special operations teams with indigenous forces and precision combat air power is a case in point.

Some commentators strongly disagree with this claim. They point either to the idiosyncratic nature of conflict in both Afghanistan and Iraq, or conversely to the fact that operations in those theaters are simply typical twentieth-century wars relying on the basic principles of firepower and maneuver (Biddle 2002). In the wider context of contemporary conflict, however, there are two important lessons to consider. First, that SOF are able to use their capabilities in high-risk environments and in response to security issues much wider than just major war. Second, the possibility of military force now being used in a domestic environment, or in non-warlike roles, is no longer a material or ideational barrier to policy makers. SOF have now truly moved from being a marginal, though at times important, component of conventional military strategy to being a central and vital element of any warfighting or security calculus.

The second significant implication of this transformation is the successful adoption of joint structures, culture and doctrine by special operations communities (at least those of Australia and the US as argued here). Such innovations may provide conventional forces with a template for the further development of joint warfighting. Innovation might also provide a working example of the benefits of a network-centric warfare approach, both in the technology used and through the tactics that SOF have developed and employed in combat. Thus, in addition to conducting operations, these forces also seem to function as a live "battle lab" for future concepts and technologies that can later be employed on a wider scale.

The danger in all this for the special operations community is that their strategic appeal and operational success have, in a way, been too good. Policy planners, particularly in the US, are already calling for large numbers of the conventional forces to become "more SOF-like," while existing SOF ought to be given sole responsibility for the prosecution of all indirect and intrastate wars. These concepts may seem like a strategic and defense-planning ideal. However, the reality is that special operations force numbers remain small, even in large armed forces such as the US, primarily because it takes many years, and significant financial and physical resources, to select and train individual operators and small units to the necessary standard, and then to hone those skills on combat operations. Any forced expansion of special operations units beyond their natural capacity to train and retain members – or indeed any significant organizational convergence with conventional units – could be counterproductive. That type of approach could result in a loss of capability and a diminution of the culture that, paradoxically, makes these forces "special." The problem of size presents defense planers and SOF themselves with a significant future policy challenge: to ensure that future forces are large enough to meet the growing demand for their capabilities, and yet small enough to maintain their high standards of training, readiness and operational performance (Fitzsimmons 2003: 216).

Note

1 The author would like to thank Michael Evans of the Land Warfare Studies Centre and Bernard Loo of the Institute of Defense and Strategic Studies for their guidance and comments on earlier versions of this chapter. The author gratefully thanks the Director of theLand Warfare Studies Centre for his permission to re-print this chapter in the present volume.

References

Ackerman, R.K. (2003) "Special Operations Forces Become Network-centric," *Signal* 57(12): 17–21.

Alberts, D., Garstka, J., and Stein, F. (2000) *Network Centric Warfare: Developing and Leveraging Information Superiority*, Washington, DC: Command and Control Research Program.

Biddle, Steven (2002) *Afghanistan and Future of Warfare: Implications for Army and Defense Policy*, Carlise, PA: US Army War College Strategic Studies Institute.

Billingslea, Marshall (2003) "Waging the War on Terrorism," speech to the Heritage Foundation. Online, available at www.heritage.org/Research/NationalSecurity/ wm256.cfm.

Bolkom, C., Deserisy, L., and Kapp, L. (2003) "Homeland Security: Establishment and Implementation of Northern Command," *CRS Report for Congress*, Washington, DC: Library of Congress Congressional Research Service.

Bostock, I. (2003) "Australian Forces go SCUD Hunting in Western Iraq," *Jane's Intelligence Review* 15(7): 20–22.

Bradley, C.M. (2004) *Intelligence, Surveillance and Reconnaissance in Support of Operation Iraqi Freedom: Challenges for Rapid Manoeuvres and Joint C4ISR Integration and Interoperability*, Newport, RI: Joint Military Operation Department, Naval War College.

Call, C.D. (2003) *US Army Special Forces Operational Interoperability with the US Army's Objective Force: The future of Special Forces Liaison and Coordination Elements*, Ft Leavenworth, KS: School of Advanced Military Studies, US Army Command and General Staff College.

Celeski, J.D. (2002) "A History of SF Operations in Somalia: 1992–1995," *Special Warfare* 2(15): 16–27.

Conford, P. and Malkin, B. (2003) "Seized: Ship they Hunted for Days," *Sydney Morning Herald* 21 April 2003.

Conner, K. (2000) *Ghost Force – the Secret History of the SAS*, London: Orion Books.

Cook, N. (2002) "Military Priorities and Future Warfare: Revolutionary Thinking," *Janes Defense Weekly* (38(11): 34–46.

Cordesman, Anthony (1999) *The Lessons and Non-lessons of the Air and Missile Campaign in Kosovo*, Washington, DC: Center for Strategic and International Studies.

Cordesman, Anthony (2003) *The Iraq War: Strategy, Tactics, and Military Lessons*, Washington, DC: CSIS Press.

Corera, G. (2003) "Special Operations Forces Take Care of War on Terror," *Janes Intelligence Review* (15(1): 42–47.

Department of Defense (1993) *Joint Special Operations Targeting and Mission Planning Procedures*, Joint Publication 3–05.5, Washington, DC: Department of Defense.

Department of Defense (2001) *Network Centric Warfare: Department of Defense Report to Congress*, Washington, DC: Department of Defense.

Department of Defense (2004) *Special Operations Force Posture Statement 2003/2004*, Washington, DC: Pentagon.

Dickson, K.D. (2001) "The New Asymmetry: Unconventional Warfare and Army Special Forces," *Special Warfare* 14(4): 14–19.

Erckenbrack, A. (2002) "Transformation: Roles and Missions of ARSOF," *Special Warfare* 15(4): 2–8.

Feickert, A. (2003) "US Special Operations Forces: Background and Issues for Congress," *CRS Report for Congress*, Washington, DC: Library of Congress Congressional Research Service.

Feickert, A. (2006) "US Special Operations Forces: Background and Issues for Congress," *CRS Report for Congress*, Washington, DC: Library of Congress Congressional Research Service.

Findlay, M., Green, R., and Braganca, E. (2003) "SOF on the Contemporary Battlefield," *Military Review* 83(3): 8–14.

Finn, P. (2001) "Wounded Army Captain Details Offensive Against Taliban," *Washington Post* 11 December 2001.

Fitzsimmons, M. (2003) "The Importance of Being Special: Planning for the Future of US Special Operations Forces," *Defense and Security Analysis* 19(3): 203–218.

Foot, M.R.D. (1970) "Special Operations, I," in E. Elliott-Bateman (ed.), *The Fourth Dimension Resistance*, Manchester: Manchester University Press.

Forero, J. (2004) "Columbian Rebel's Capture was Result of Hunt Aided by US," *New York Times* 4 January 2004.

Gagnon, G. (2002) "Network-centric Special Operations: Exploring New Operational Paradigms," *Air and Space Power Chronicles* 4 February 2002.

Gellman, B. and Linzer, D. (2004) "Afghanistan, Iraq: Two Wars Collide," *Washington Post* 22 October 2004.

Gilmore, G. (2003) "Special Operations: Force Multiplier in Anti-terror War," *American Forces Information Service* 13 November 2003.

Gray, Colin S. (1996) *Explorations in Strategy*, Westport, CT: Greenwood Press.

Gray, Colin S. (1999) *Modern Strategy*, New York: Oxford University Press.

Guttieri, K. (2003) "Homeland Security and US Civil–Military Relations," *Strategic Insight*, US Naval Postgraduate School Newsletter, 1 August 2003

Hersh, Seymour (2003) "Moving Targets," *New Yorker* 15 December 2003.

Hester, J.L. (2003) *Integration of Special Operations Forces into the Joint Targeting Process*, Ft Leavenworth, KS: US Army Command and General Staff College.

Horner, David M. (2002) *SAS: Phantoms of War*, Sydney: Allen & Unwin.

Jackson, S.A. (2003) *Tactical Integration of Special Operations and Conventional Forces Command and Control Functions*, Ft Leavenworth, KS: School of Advanced Military Studies, UA Army Command and General Staff College.

Jones, C. (2002) "Transnational Threats and the Role of the Military in the 21st Century," paper presented at the Australian Defense Studies Centre Homeland Security Conference, Canberra, 1 November.

Jones, G.M. and Tone, C. (1999) "Unconventional Warfare: Core Purpose of Special Forces," *Special Warfare* 12(4): 1–13.

Kennedy, H. (2002) "Will Special Ops Success Change the Face of War?" *National Defense Magazine* 86(579): 20–21.

Lewis, D. (2003a) "Guarding Australians Against Terrorism," *Australian Army Journal* 1(2): 45–52.

Lewis, D. (2003b) "Inside and Outside the Battlespace: Understanding the Rise of Special Operations in Australia," *Australian Army Journal* 1(2): 53–58.

Luttwak, Edward (1982) *A Systematic Review of "Commando" (Special) Operations 1939–1980*, Potomac, MD: C & L Associates.

Minister for Defense [Australia] (2002) "Counter-terrorism Capabilities Doubled," *Media Release 204/02*, 14 May 2002.

Minister for Defense [Australia] (2003) "New Special Operations Command," *Media Release 47/2003*, 5 May 2003.

Minister for Defense [Australia] (2004) "New Commander Special Operations Announced," *Media Release 181/2004*, 14 October 2004.

Mitchell, Mark E. (1999) "Strategic Leverage: Information Operations and Special Operations Forces," Monterey, CA: US Naval Postgraduate School.

Moreman, T.R. (1992) "The British and Indian Armies and North-West Frontier Warfare: 1849–1914," *Journal of Imperial and Commonwealth History* 20(1): 35–64.

Nardulli, B. (2003) *The Global War on Terrorism: An Early Look at Implications for the Army*, Arlington, VA: RAND Corporation.

Newman, D. (2005) "Operation 'White Star': a UW operation against an insurgency," *Special Warfare* 17(4): 28–36.

Newman, R.J. (1998) "Hunting War Criminals," *World Report* 6 July 1998.

Noonan, M.P. and Lewis, M.R. (2003) "Conquering the Elements: Thoughts on Joint Force (Re)Organization," *Parameters* 33(3): 31–45.

Ripley, T. (2003) "Iraq's Western Desert a Special Forces Playground," *Jane's Defense Weekly* 9 April 2003.

Robinson, Linda (2005) "Plan of Attack," *US News and World Report* 8 January 2005.

Schoomaker, P.J. (1998) "US Special Operations Forces: The Way Ahead," *Special Warfare* 11(4): 2–9.

Shanker, T. and Schmitt, E. (2003) "Pentagon Says a Covert Force Hunts Hussein," *New York Times* 7 November 2003.

Sullivan, D.M. (2002) "Transforming America's Military: Integrating Unconventional Ground Forces into Cobat air Operations," Newport, RI: US Naval War College, Joint Maritime Operations Department.

Tugwell, M. and Charters, D. (1984) "Special Operations and the Threats to United States Interests in the 1980s," in Barnett, F.R., Hugh Tovar, B., and Shultz, R.H. (eds) *Special Operations in US Strategy*, Washington, DC: National Defense University Press.

Valenzuela, A.A. and Rosello, V.M. (2004) "The War on Drugs and Terrorism: El Salvador and Colombia," *Military Review* 84(2): 28 35.

6   Unmanned aerial vehicles

Missions, challenges, and strategic implications for small and medium powers

Manjeet Singh Pardesi

The absorption of modern info-communications technologies (ICT) has transformed the United States (US) military. This transformation has in turn changed the conduct of warfare in two ways – by enhancing platforms and by enabling network-centric warfare. Unmanned aerial vehicles (UAV), armed and unarmed, are playing a crucial role in this revolution, as they provide the military with a new platform that exploits the advances in ICT, and at the same time are integral to the network-centric warfare concept. Although interest in UAV is as old as the history of manned aviation, UAV started to demonstrate their military effectiveness only in recent conflicts such as Afghanistan in 2001, and Iraq in 2003. The former is of particular interest because the coalition campaign against the Taliban witnessed UAV attacking targets in addition to their traditional mission of intelligence gathering and guiding weapons to their target.

This chapter examines one key question: do UAV represent a truly disruptive technology? Will UAV have an impact on the future of manned aircraft, and how does the increased use of unmanned platforms alter the strategic landscape? To this end, this chapter will examine various air operations – intelligence, surveillance, and reconnaissance (ISR), suppression of enemy air defenses (SEAD), and counter-air – to establish the transformative impact of UAV, if any. This chapter will also briefly discuss how mini/micro aerial vehicles (MAV), a subset of UAV, are likely to be deployed on the battlefield. This chapter will go on to highlight the strategic implications of UAV for small and medium powers. It will highlight the missions that are likely to be transformed with the introduction of UAV and make a policy recommendation to these states with limited defense (and R&D) resources to invest only in these mission areas.

History of unmanned vehicles

Even though the concept of unmanned vehicles seems somewhat revolutionary in nature, it is not modern technology (Ehrhard 2000: 653–6). Starting with the kite, the Chinese sought to achieve military advantage with devices held aloft by aerodynamic forces, as early as third century BC. In Europe, the first recorded use of kites for military purposes was in the Battle of Hastings in 1066. The Americans first used unmanned aviation in combat when journalist William

Eddy took hundreds of photographs during the Spanish–American War in 1898 from cameras suspended from a kite. Samuel Pierpoint Langley attained the first heavier-than-air, sustained (lasting over one minute), powered flight with a steam-powered aircraft over the Potomac River on 6 May 1896 (O'Reilly 1999: 9–10). But after the Wright brothers' first piloted, powered flight at Kitty Hawk on 17 December 1903, unmanned aviation had to take a backseat to manned aviation.

As early as World War I, the US began to look into technologies that pre-saged contemporary cruise missiles (Clark 2000: 7–12). It was really during the inter-war years, as radio technologies came of age – a critical technological development for unmanned aviation technologies – that eventually resulted in the first successful prototypes of UAV platforms. But it was the downing of the U-2 reconnaissance aircraft piloted by Francis Gary Powers and an RB-47 over the Barents Sea in 1960 that intensified the search for workable unmanned vehicle technologies (Clark 2000: 12–23). The US Air Force (USAF) awarded Ryan Aeronautical two classified contracts to demonstrate its drones for photo-reconnaissance missions, which eventually led to the development of the Ryan-147 Lightning Bug, which flew a total of 3,435 operational reconnaissance UAV sorties throughout the Vietnam War. In 1971, the US recorded another first, when a modified Lightning Bug BGM-34A successfully engaged and destroyed a guided air-to-surface missile against a simulated SAM (surface-to-air missile) site. Since this system was still under development, however, unmanned combat aerial vehicles (UCAV) played no part in the Vietnam conflict. By 1979, defense budget cuts forced the various UAV/UCAV programs to be shelved or cancelled.

Successful Israeli UAV missions during military operations in Lebanon in 1982 rekindled US interest in this technology (Bone and Bolkcom 2003: 2). The US Navy (USN) acquired the Pioneer UAV from Israel, which provided tactical-level intelligence during Operation Desert Storm in 1991. The ineffectiveness of the Tomahawk Land-Attack Missile (TLAM) attacks on Osama bin Laden's camps in Afghanistan, in retaliation for al-Qaeda-sponsored bombings of US embassies in Africa in August 1998 further encouraged US military interest in new roles for armed and unarmed UAV (Gormley 2003: 416). During Operation Allied Force, UAV performed numerous functions, including target identification and acquisition, probing of Serbian air defenses, monitoring Serbian ethnic cleansing operations, bomb damage assessment (BDA), electronic intelligence operations, airborne communication relays and jamming of Serbian communications.

The latest developments in the history of unmanned vehicles can be found in the recent conflicts in Afghanistan and Iraq. In Afghanistan, the Predator UAV started performing armed reconnaissance missions as mentioned earlier and the Global Hawk UAV made its debut in the skies over Afghanistan in 2001, even though it was still an experimental system then. In the recent war in Iraq, Global Hawks provided imagery of Iraqi Republican Guard divisions, while the Predators continued their combat role by attacking high-value targets in Iraq. Surveillance UAV also assisted US Special Forces in their mission of preventing the Iraqis from launching any hidden Scud missiles (Krepinevich 2003).

Roles and missions

While there is a good deal of confidence in the underpinning technology of unmanned platforms, there is also a good deal of uncertainty surrounding their roles and missions. UAV/UCAV are likely to play a key role in mission areas commonly categorized as "the dull, the dirty and the dangerous" (Office of the Secretary of Defense 2002: iv). This section discusses three key air missions (ISR, strike/SEAD, and counter-air) to determine if UAV/UCAV can replace manned platforms in some or all of these roles. This will be followed by a short analysis on the role of MAV on the battlefield. It must be pointed out that the move towards unmanned platforms is not necessarily due to the inadequacy of manned aircraft. Rapid technological advancement over the past decade has led to a "technological push" in this direction. Moreover, since the end of the Cold War, the US has been attempting to replace manpower with technology, mostly because it retains strategic interests in every corner of the globe but is increasingly hesitant to commit its military personnel for several of these missions. The greater emphasis placed on unmanned platform technologies is a result of all these developments.

Intelligence, surveillance, and reconnaissance (ISR)

UAV have been traditionally used as ISR assets, and their ability to do so is being boosted by the advances in sensor and modern ICT. For the US, ISR collection is a critical factor in achieving the Joint Vision 2020 operational concept of precision engagement (Department of Defense 2000: 12). Furthermore, the growth of asymmetric warfare only further accentuates the importance of unmanned ISR platforms.

During the Vietnam War, the photos provided by the Ryan-147 Lightning Bug revealed precise locations of SAM sites, enemy airfields, ship activity in Haiphong Harbor and BDA, provided intelligence that otherwise would have been obtained only if manned aircraft were sent in harm's way (Clark 2000: 5–16). In Operation Desert Storm, the Pioneer UAV played a critical role in target designation, damage assessment and reconnaissance that contributed to the tactical successes of coalition forces (Clark 2000: 34–5). In Afghanistan, the Global Hawk UAV was used for reconnaissance prior to the strikes and for post-strike BDA (Persinos 2002). The Predator UAV was used in Afghanistan to feed imagery to AC-130 special operations gunships and special operations teams on the ground (Bone and Bolkcom 2003: 14). Global Hawk accounted for only 5 percent of intelligence sorties during Operation Iraqi Freedom but produced 50 percent of the information on time-sensitive targets (Donnelly and Vickers 2003). It is important to note that unmanned aerial vehicles retreated to their traditional role of reconnaissance in Iraq in spite of some successes in combat role in Afghanistan. In Afghanistan, barely a dozen UAV launched 115 Hellfire air-to-surface (ASM) missiles and laser-designated 525 targets. But in Iraq, where more than 56 larger UAV and more than 60 smaller portable ones were used, UAV launched only 62 Hellfire ASM and designated only 146 targets. The main reasons were winds and sandstorms in the Iraqi desert (the fact that these aircraft are much lighter than their manned counterparts made them more vulnerable to strong winds) and the increased need for intelligence in the Iraqi campaign (Kaufman 2003).

However, UAV face two competing platforms in ISR missions – manned platforms and satellites. While providing significant improvements in information collection capabilities over these competing systems, UAV also pose some serious limitations.

Airborne Warning and Control System (AWACS) and Joint Surveillance Target Attack Radar System (JSTARS) aircraft, being large manned platforms, have limited maneuvrability and defenses against enemy missile attacks. Manned platforms provide high resolution data and are extremely flexible at adapting to multiple mission scenarios, however, their main limitation is their loiter time. UAV on the other hand are capable of long loiter times; are smaller and hence stealthier than manned platforms; much less costly to procure, operate, and support; and avoid putting pilots at risk. Unlike the UAV, the loss of even one of these expensive manned systems is likely to cause severe domestic political repercussions for the US, UAV platforms would therefore seem to be natural replacements for these manned platforms. However, fast jet-based tactical reconnaissance remains a much sought after, but scarce capability for UAV. It is possible that in the future UAV will be faster and more maneuvrable, but it must be remembered that higher speed creates penalties for loiter time, one of the biggest assets of unmanned platforms.

Furthermore, given the current state of technology, UAV cannot completely replace AWACS and JSTARS in ISR missions. Advanced sensor technology is still under development and information technologies (IT) are not sufficiently developed to perform the battle management and command and control functions currently handled by AWACS and JSTARS. Furthermore, due to their inability to absorb data and reason (at least for the foreseeable future), UAV cannot process and relay the same amount of data as a pilot in the cockpit (who can do so by learning, experiencing, and by intuition) and cannot maintain a 360° situational awareness. The US military is seeking to further develop technologies that can provide sensors with high definition television (HDTV) standards, foliage penetration radar (FOPEN) with hyperspectral imagery, synthetic aperture radar (SAR) and moving target indication (MTI) mode to track targets in all types of terrain throughout the spectrum of military operations (Hewish 2002). Nevertheless, it is worth noting that it was the integrated use of Global Hawk, Predator and JSTARS systems that was the key factor behind the shattering of the Republican Guard and the success of the SCUD suppression campaign in western Iraq during the war in Iraq (Donnelly and Vickers 2003).

Operation Desert Storm highlighted the pivotal role that satellites will have in future conflicts, as space (due to the salience of satellites) became an area of strategic significance. In comparison to satellites, however, the UAV has a major advantage, in that it is easier to alter its flight path and coverage. Moreover, being cheaper than satellites, they provide a comparatively cost-effective method of ISR collection. UAV also have an additional advantage of being able to fly closer to the target (O'Hanlon 2000: 34). However, the major drawback with UAV as mentioned above is their lack of situational awareness. This short-coming can be overcome by integrating UAV with reconnaissance satellites. However, such real-time interactive command and control systems demand very large bandwidths to facilitate flight controls, video reception and transmissions, and in this regard, UAV are major consumers of bandwidth (Lewis 2002: 44–6). Since 11 September 2001, the need has increased eight-fold due to the war in Afghanistan and the Global War On Terror (Bone and Bolkcom 2003: 17–18). Locating the mission control on a standoff aircraft (within line-of-sight) would decrease the dependency on satellites generated by stationing the mission control on the ground thousands of miles away. More autonomous UAV will also require less bandwidth as more data will be processed on board. It should also be remembered that since the UAV fly in close proximity to the target, they will also need to have a high signal-to-noise ratio (especially if they are flying far from their control station), thus increasing their possibility of detection.

The way forward is thus to integrate manned, unmanned, and satellite-based sensors to create a common operational picture of the battlefield. However, the spread of unintegrated information can be potentially disastrous in a military campaign. Development of ICT and software algorithms to fuse the data pro-vided by the three platforms will be crucial to ISR operations in the future. A successful ISR mission must have a reliable, robust, secure and high-capacity communications infrastructure. The information collection system of the future is likely to be based on space-based assets providing wide area surveillance at a low level of resolution, but looking for cues that require detailed monitoring, which can then be performed by manned and unmanned vehicles.

Strike and suppression of enemy air defenses

After the embassy bombings in Africa, the US military response focused on tar-geting Osama bin Laden and his training camps in Afghanistan with TLAM. This strategy managed to keep US troops out of harm, but it suffered from many operational limitations, the most important being the long delay between acquir-ing reliable intelligence on the precise location of time-sensitive targets (from the skies over Afghanistan) and the execution of an actual cruise missile attack (from ships in the Arabian Sea). The US was looking for an armed reconnais-sance platform that could strike time-sensitive targets. One technological solu-tion to this problem was for the USAF to fit Hellfire-C ASM to a UAV, in the process turning it into a weapons platform. There have been a number of instances – the killing of Muhammad Atef, al-Qaeda's chief of military opera-tions on 15 November 2001; and the killing of a top al-Qaeda operative, Ali

Qaed Sinan al-Harthi on 3 November 2002 – when armed UAV platforms have successfully attacked and destroyed time-sensitive targets (Gormley 2003: 416–17). Such incidents demonstrate the potential usefulness of armed UAV in a possible new role for the armed UAV – suppression of enemy air defenses (SEAD).

SEAD is an important mission as it helps in attaining air superiority. The air forces can attack the heart of the enemy (i.e., perform the interdiction mission) only after gaining command of the air. The Predator UAV was credited with two strikes in Operation Iraqi Freedom in March 2003 – one strike was against an anti-aircraft platform while the other was against a television satellite dish in Baghdad (Bone and Bolkcom 2003: 14). The US is currently developing a new version of the armed Predator UAV, called Predator B, which will have the capability to carry eight Hellfire ASM instead of two. The US is also developing newer platforms – UCAV – that are being developed with a primary offensive mission of strike and SEAD.

To determine the efficacy of UAV in a SEAD role, the US will need to consider two rival challenges – the adoption of new counter-tactics by its opponents, and the development of new air defense systems. The US currently relies exclusively on the USAF F-16 and the USN EA-6B for defense suppression missions. The loss of a modern, expensive platform like the F-16 (and its pilot) will be a major political embarrassment for the US During Operation Desert Storm, the F-117 allowed the US to engage in parallel warfare; i.e., it freed the US from having to first roll back enemy air defenses, and enabled the F-117 aircraft to attack targets right at the heart of the enemy within the opening minutes of the conflict (Keaney and Cohen 1995: 189–93). By the time of the NATO operations over Kosovo, however, the Serbs had learnt "shoot and scoot" tactics to overcome similar parallel operations by NATO forces. This meant abandoning a fixed air defense system, relying instead on purely mobile SAM systems as the main Serbian air defenses. This allowed the Serbs to launch 700 missiles in the course of the 78-day conflict and was a source of enormous frustration to NATO (Jumper 2001: 27).

Even without innovations in such tactics, US airpower is likely to face anti-aircraft defenses such as SAM and other advanced air defense systems that will seriously undermine its ability to bring decisive airpower to bear in any conflict scenario. The range of modern SAM (estimated to be between 50 and 250 miles), is forcing the US to develop tactics and other systems to cope with this increasing threat to its airmen (Tirpak 2001). SAM launched at standoff ranges can be very difficult to detect. Combined with the missiles' high speeds and agility, this means that the targeted aircraft/UAV will have a very small probability of avoiding the SAM. Unmanned platforms have G-force tolerances ($\pm 12G$) that do not significantly exceed those of the human pilot (between $-3G$ and $+9G$) and hence are not significantly invulnerable to missiles (Erhard 2000: 574). It must be pointed out that the cost arithmetic further complicates the analysis and is not useful in determining the efficacy of UCAV over current standoff systems like cruise missiles (Chapman 2002: Sweetman 2003). JDAM deployed on UCAV may be cheap compared to the Tomahawk, but the UCAV, which is an expensive recoverable platform, is likely to suffer considerable attrition due to its proximity to the target.

Thus, while UAV/UCAV will likely play important roles in electronic attack missions, they will probably at best play a limited role only as the future use of electromagnetic pulse (EMP) and directed energy (DE) weapons will increase the risk of jamming for the unmanned platform itself. The new S-400 SAM system with a range in excess of 250 miles would also render manned standoff jamming platforms useless (Sweetman 2003: 27). However, it might be possible to use low-cost UAV and/or decoys to locate the positions of the SAM sites, as part of a more reactive approach to SEAD missions. This together with UCAV equipped with passive sensors (an extremely stealthy platform), possibly represents an effective counter to mobile defenses. However, given the primitive nature of extant target acquisition and fire control systems, a human operator will still be required to authorize the "kill," thereby increasing the bandwidth requirements; also, integration with other ISR platforms is necessary to locate time-sensitive targets. These constraints make the use of UCAV in reactive SEAD missions problematic. UCAV are more likely to play an important role in pre-emptive SEAD missions (where the exact locations of enemy SAM sites are known). UCAV, integrated with manned and unmanned assets like AWACS, F-16s, F-117s, Global Hawk, and communications satellites will play a role in future SEAD missions, especially in high threat environments; however, they will be only another of several tactical solutions to the SEAD problem.

Counter-air

In March 2003, a Predator UAV launched a Stinger air-to-air missile at an Iraqi MiG before the Iraqi aircraft shot it down (Fulgham 2002: Fulgham 2003). This has led to the speculation that armed UAV/UCAV will play a role in counter-air operations (and by extension as air superiority fighters in the future). During Operation Desert Storm, coalition air forces flew over 13,000 counter-air missions, averaging 340 sorties daily, employing such traditional manned air platforms as the F-14A/D, F-15C, and F/A-18 aircraft (Keaney and Cohen 1993: 56). The same air assets were applied in similar missions during Operation Allied Force.

Stealth, maneuvrability, and cost are the most important design pre-requisites for air superiority fighters of the future. This means that the Lockheed Martin F-22 Raptor will likely play the key role in America's air superiority efforts in the years ahead. Whether or not a UCAV will replace manned platforms is a crucial question as American air superiority in a future conflict depends on the answer to this question. This is also a timely question since the decisions taken today will guide the research, development, production, and training of the new system (manned or unmanned replacement of the F-22 fighter) over the next two decades (a period at the end of which the F-22 will most likely retire).

Aerial combat is the most challenging mission for manned aircraft to perform. Standoff-range air-to-air missiles (AAM) may not always kill the adversary (especially one equipped with significant counter-air assets and capabilities like the MiG-29 Fulcrum and the Su-27 Flanker), so close-in engagements may often be necessary. For this reason, combat survivability remains the most significant limitation to UAV employment (Banks 2000: 18). In a latter section, it will be noted that limitations imposed by line-of-sight data transfer requirements will enhance the role of satellite communications. However, the current US and allied satellite communications infrastructure is incapable of supporting any sizable number of UAV or UCAV. The Global Hawk UAV consumed five times the total bandwidth used by the entire US military in the Gulf (Klausner 2002). Autonomous systems will reduce bandwidth requirements; however, it is unlikely that the UCAV will replace the manned aircraft in all operations as some politically sensitive targets will still need a human operator to make the "kill decision." Moreover, cognitive systems based on artificial intelligence (AI) are unlikely to replace the human completely, even though significant developments are likely to occur over the next two decades. Finally, stealth requirements dictate that the UCAV weapons be small and, by extension, precise. The weaponization of the unmanned platform for air superiority missions is not likely to happen over the next two decades (Lewis 2002: 50–2). In the near future, the UCAV will likely deploy with extant AAM like Sidewinder or AMRAAM (advanced medium-range air-to-air missiles). UAV will be used predominantly to provide active sensors against highly lethal anti-aircraft weapons in support of manned vehicles.

Mini/micro-aerial vehicles

The US is also heavily investing in a new class of unmanned platforms – Mini/Micro Aerial Vehicles (MAV). MAV are a type of UAV, only much smaller, some as small as six inches. These compact lightweight air vehicles carrying miniature sensors are currently playing a key role in the war against terrorism (Dornheim and Taverna 2002: Hewish 2002). While MAV are more vulnerable to attack and loss due to their low altitude, this is compensated by the fact that they are extremely stealthy and very cheap. Their compact size and low weight will allow them to be carried by individual soldiers. The USAF is currently deploying MAV for force protection in the shape of the Lockheed Martin SentryEye.

MAV can have tremendous potential for ISR operations, given their inherent flexibility. In the battlefield, individual infantrymen can operate them for local reconnaissance. MAV integrated with a high-flying UAV offers the potential for circumventing the need to develop foliage penetration sensors. They can also play an important role in urban operations where stealthy airborne assets closer to the ground may be required. In the sea, MAV can also be deployed from ships to gather intelligence in order to prevent acts of maritime terrorism. They may also be fielded in a hostile environment to detect people with shoulder-fired missiles to attack aircraft. MAV can also play an important role in real-time detection and analysis of a biological or a chemical agent in an infected environment. They are also likely to play an important role in humanitarian missions, e.g., searching for survivors amidst rubble from earthquakes.

Swarms of MAV equipped with sensors and miniaturized warheads are also theoretically capable of attacking high-value targets such as radars and launchers of SAM sites, i.e., they are likely to play an important role in SEAD missions in the future (Hewish 2002). Global Positioning System (GPS) allows precise autonomous navigation and position reporting for MAV, which are critical to the military application of these technologies. Of course, MAV are limited by their short operating ranges and high damage potential (especially due to the prevailing weather).

Evaluation

Fiscal challenges

The US DoD is planning to invest around $10 billion in UAV technologies in the first decade of this century and plans to quadruple today's 90-aircraft inventory by then (Fulgham 2003a). This invites comparison with the fact that the USAF has spent close to $20 billion on the F-22 air superiority fighter which will cost at least $100 million per aircraft to produce and will purchase close to 300 F-22s (Fallows 2002). The US will also spend between $28 million to $38 million per aircraft on a new tactical fighter called the Joint Strike Fighter (JSF) and with the intention to purchase up to as many as 3,000 JSFs. The total system cost of the Predator UAV, which is about $28.3 million, is about the same as a single seat F-16A (Mustin 2002). Although the unmanned platform might be cheaper than its manned counterpart, the UAV system on the whole is not always less expensive. Besides, it is estimated that the DARPA/Boeing X-45 UCAV will cost about $25 million each (Sweetman 2003).

It is clear from Table 6.1 that the unmanned platform does not necessarily offer the cost-effectiveness that it promises. Unmanned systems are "attritable," but not expendable, i.e., it is fine to lose them when the alternative is losing a manned platform. However, these are not cheap systems; thus they are not entirely expendable either. It should also be highlighted that UAV are on average lost at a much higher rate than manned aircraft (Sweetman 2003). For the UAV to replace the manned platform, it must offer either the same or even

Table 6.1 Approximate costs of current and future manned and unmanned air platforms

Manned/unmanned system	Cost
F-22 Raptor (per unit)	US$100 million
JSF (per unit)	US$28–38 million (depending upon specs.)
Predator System	~US$28 million
X-45 UCAV (per unit)	~US$25 million

higher level of reliability as the manned platform. DARPA, Boeing, and the US military are working together to develop a pure UCAV called the X-45 and its naval version UCAV-N, which are likely to play a major role in strike missions and electronic attacks in the future. These systems are likely to become operational in the 2008–2015 timeframe. Nevertheless, it is important to remember that these systems are still under development and would need to undergo extensive testing to prove their technological capability. Unless they are tried and tested in an actual operation (in small numbers at first), these systems are unlikely to challenge the manned platform in any significant way.

Strategic implications for small and medium powers

It is obvious that UAV technologies do provide the US with the policy option of intervening militarily anywhere in the world whenever its interests are threatened without necessarily putting its forces in harm's way. That being said, there is nothing to stop other states from themselves acquiring UAV technologies and/or weapons of mass destruction (WMD) to oppose a US-led intervention (Gormley 2003: 410). The operational implications of this are as yet indeterminate. Possibly, and somewhat ironically, the greatest risk to US national security right now is the prospect of terrorists using armed UAV to conduct attacks against the US (Gormley 2003: 413). In the US missile defense system, UAV are also likely to play an important role as interceptors to destroy ballistic missiles. UAV technologies are clearly a potential double-edged sword for the US. The proliferation of armed UAV in its opponents' arsenals is almost certainly going to complicate the cost-per-kill arithmetic for US missile defenses (Gormley 2003: 411). UAV will also enable regional powers to bolster their power projection capabilities. India has raised its profile in the Indian Ocean Region by operationalizing its first full-fledged UAV base in Kochi where its Southern Naval Command is based. India also plans to set up UAV bases in Port Blair in the Andamans and Lakshadweep islands (Joseph 2001).

What lessons should small and medium powers draw from the current and projected technological challenges and operational capabilities of UAV/UCAV? The extent to which small and medium powers can absorb high technology into their militaries depends on various factors. Any country with a well-educated population and workforce, knowledge-based economy, a sophisticated defense–industrial base, political stability and extensive ties with Western companies is well placed to absorb advanced technologies into its military forces.

It is in ISR missions where the use of UAV is the most promising and it is precisely here where small and medium powers are most likely to spend their resources. ISR capability will enable these states to gain dominant battlespace knowledge in any conflict. Of course there is the important obstacle of successfully integrating these technologies into existing armed services, which can be an equally daunting and time-consuming task (Bone and Bolkcom 2003: 13). Furthermore, technological challenges, especially bandwidth requirements and systems integration, pose considerable hurdles. These states will need to invest in satellite technology to solve the problems posed by bandwidth limitations. UAV are also likely to have numerous commercial applications that will interest small and medium powers. This is good news for the defense–industrial sector, as it is likely that some of the research on unmanned technologies will be carried out by the commercial–university sector. Furthermore, many of the technologies involved (autopilot systems, satellite navigation and guidance systems, digital mapping technologies for mission planning, and collision avoidance systems, etc.) are dual-use technologies.

UCAV technology is still in its infancy and has not yet been conclusively demonstrated on the battlefield. The US is perhaps the only country with enough resources to expend on this unproven (and thus far, unexperimented) technology. The high costs involved in experimenting with this immature technology will likely make this technology area prohibitive for small and medium powers, at least for the immediate future. After all, research and development costs continue to approximately double those of procurement (Bone and Bolkcom 2003: 11). Small and medium powers have to recognize that their more limited resources will likely prohibit an across-the-board approach to UAV technologies that the US might hypothetically adopt, although the experience of a small power like Singapore might provide some signal lessons. Singapore is working on a naval surveillance UAV named LALEE (low-altitude, long-enduring endurance) with the European Aeronautic Defense and Space Company (EADS). EADS has apparently expressed interest in the prospects of other collaborative ventures with Singapore. LALEE has also attracted interest from the US, France and Sweden.

Indeed, in spite of their size and budget constraints, small and medium powers will probably not be deterred by the high costs of UAV technologies. These unmanned platforms are attractive precisely because they can fulfill important functions like battlefield surveillance and armed reconnaissance. MAV technologies offer the potential to substantially transform urban operations and special operations; small wonder that small and medium powers have manifested such interests in these unmanned technologies. Furthermore, UAV will also play a key homeland security role for these states. In addition, collaboration (especially in the area of research and development), licensed production and joint marketing are other areas that will allow small and medium powers to enter into this emerging technology area.

Conclusion

The unmanned aerial vehicle is an innovative weapon system that avoids putting a pilot in harm's way, but it is not a truly disruptive technology as there will always be missions that will require the manned aircraft. Likewise, the unmanned platform has less flexibility, cannot analyze its environment, and remain vulnerable to counter-measures and destruction by anti-aircraft fire. It is clear that many advanced unmanned platforms are as expensive as manned aircraft and their high cost makes them attritable, not expendable. Their software complexity, automation and communications architecture makes them operationally unreliable for many missions. Thus far, communications technology has limited the effectiveness of the unmanned platform, especially its armed counterpart. Finally, UAV also face considerable challenges from competing systems. Satellites not only provide better situational awareness, but also avoid international norms for violating national/sovereign airspace and are thus far invulnerable to being shot down. Standoff-range cruise missiles like TLAM have thus far proven superior in weapon delivery roles. However, many dull, dirty and dangerous missions will see an increased role for the unmanned platform.

Nevertheless, unmanned technologies will likely perform the critical ISR missions in future military operations, along with tactical missions together with their manned counterparts. Pre-emptive strike and SEAD missions will also likely be increasingly dominated by unmanned platforms, but reactive SEAD missions will likely remain the domain of manned platforms due to the proliferation of sophisticated anti-aircraft defenses. UAV are also likely to play an important but limited role in electronic attack missions, although the proliferation of sophisticated counter-air assets makes them unsuitable for counter-air missions and communications and automotive technology limitations together with political ones (the authorization to fire) reduces their usefulness for combat missions. It is unlikely for the unmanned platform to make significant inroads into force application roles in the immediate future.

However, it is important to remember that unmanned platforms will probably never totally replace the manned platform, if only because the manned platform, by virtue of its human operator, provides a sense of situational awareness and understanding that the unmanned platform will likely never achieve. Even artificial intelligence systems can at best only improve existing technology; they can never supplant the human under the uncertainties and rapid changes of war.

References

Banks, Ronald L. (2000) *The Integration of Unmanned Aerial Vehicles into the Function of Counterair*, dissertation, Maxwell Air Force Base, Alabama: Air Command and Staff College, Air University.

Bone, Elizabeth and Bolkcom, Christopher (2003) *Unmanned Aerial Vehicles: Background and Issues for Congress*, Washington, DC: Congressional Research Service.

Chapman, Robert E., II (2002) "Unmanned Combat Aerial Vehicles – Dawn of a New Age?" *Aerospace Power Journal*, 16(2): 60–73.

Clark, Richard M. (2000) "Uninhabited Combat Aerial Vehicles – Airpower by the People, For the People, But Not With the People," *CADRE Paper No. 8*, Alabama: College of Aerospace Doctrine Research and Education, Air University, Maxwell Air Force Base.

Department of Defense (2000) *Joint Vision 2020*, Washington, DC: US Government Printing Office.

Donnelly, Thomas and Vickers, Michael (2003) *Iraq: Lessons Learned*. Online, available at: www.aei.org/events/filter.,eventID.337/summary.asp.

Dornheim, Michael and Taverna, Michael (2002) "War on Terrorism Boosts Deployment of Mini-UAVs," *Aviation Week & Space Technology*, 8 July.

Ehrhard, Thomas P. (2000) *Unmanned Aerial Vehicles in the United States Armed Services: A Comparative Study of Weapon System Innovation*, PhD dissertation, Baltimore: Johns Hopkins University.

Fallows, James (2002) "Uncle Sam Buys an Airplane," *The Atlantic Monthly*, June 289(6): 62–74.

Gormley, Dennis M. (2003) "New Developments in Unmanned Air Vehicles and Land-attack Cruise Missiles," *SIPRI Yearbook 2003 – Armaments, Disarmament and International Security*, Oxford: Oxford University Press.

Hewish, Mark (2002) "Small, But Well Equipped," *Jane's International Defense Review*, October, vol. 35: 53–62.

Joseph, Josy (2001) "Navy to Use UAVs to Spy on Sea-lanes," *Reddif News*. Online, available at www.rediff.com/news/2003/jan/31uav.htm.

Jumper, John P. (2001) "Global Strike Task Force," *Air Power Journal*, 15(1): 24–33.

Kaufman, Gail (2003) "Shot Fewer Missiles than in Afghanistan," *Defense News* 8 December 2003: p. 29.

Keaney, Thomas A. and Cohen, Eliot A. (1993) *Revolution in Warfare? Air Power in the Persian Gulf*, Annapolis: Naval Institute Press.

Klausner, Kurt (2002) "Command and Control of the Air and Space Forces Requires Significant Attention to Bandwidth," *Air & Space Power Journal*, 16(4): 69–77.

Krepinevich, Andrew (2003) *Operation Iraqi Freedom: A First Blush Assessment*, The Center for Strategic and Budgetary Assessments (CSBA). Online, available at www.csbaonline.org/4Publications/Archive/R.20030916.Operation_Iraqi_Fr/R.20030916.Operation_Iraqi_Fr.htm.

Lewis, William K. (2002) "UCAV – The Next Generation Air Superiority Fighter," dissertation, Maxwell Air Force Base, Alabama: School of Advanced Air Power Studies, Air University.

Mustin, Jeff (2001) "Flesh and Blood: The Call for the Pilot in the Cockpit," *Air & Space Power Journal – Chronicles Online Journal*. Online at: www.airpower.au.af.mil/airchronicles/cc/mustin.html.

O'Hanlon, Michael (2000) *Technological Change and the Future of Warfare*, Washington, DC: Brookings Institution Press.

O'Reilly, Thomas (1999) *Uninhabited Air Vehicle – Critical Leverage System for our Nation's Defense in 2025*, Alabama: Air Command and Staff College, Air University, Maxwell Air Force Base.

Office of the Secretary of Defense (2002) *Unmanned Aerial Vehicles Roadmap 2002–2027*, Washington, DC: Department of Defense.

Persinos, John 2002 "Unmanned Aerial Vehicles: On the Rise," *Aviation Today*, February 2002

Sweetman, Bill (2003) "UCAVs Grow Fat on Requirements," *Jane's International Defense Review*, vol. 36: 44–9.

Tirpak, John A. (2001) "The Double Digit SAMs," *Air Force Magazine*, 84(6): 48–9.

7 The RMA and "military operations other than war"

A swift sword that cuts both ways

David J. Betz

In the early 1980s when the Soviet General Staff began to ponder the shape of future of wars what they saw worried them: a military–technical revolution (MTR) loomed in which long-range precision munitions, enhanced sensors and electronic-control systems promised to turn conventional weapons, by dint of very significant increases in their destructive power, into weapons of mass destruction in terms of the former's strategic effect. In the Soviet viewpoint, this MTR worked to the advantage of the West because of the latter's strength in computing and micro-electronics, and threatened Soviet strengths – its vast armoured forces – which would be destroyed by long-range precision fire in operational depth before they could make their presence felt on the frontline. In general the West was slow to recognise the scope of the changes afoot (Department of Defense 1982; Perry 1991). It was only in light of the spectacular performance of the United States (US) military in the 1991 Gulf War that awareness of the revolution, subsequently called the revolution in military affairs (RMA), became more general, debate about it grew more widespread, and it began to be incorporated in policy and doctrine – notably in the US in *Joint Vision 2010* (Joint Chief of Staff 1996) and *Joint Vision 2020* (Joint Chief of Staff 2000).

The point of recounting the origins of the RMA is to establish the rationale of this chapter: the concept of the RMA originated in the context of large-scale, potentially unlimited, high-intensity, force-on-force combat between two likely-armed superpowers; its promise was hinted at first in the 1991 Gulf War (and more clearly shown 12 years later in the 2003 Iraq War) in which the militarily incompetent Saddam Hussein obligingly fought in a manner which maximised what Stephen Biddle (1996) called the "synergistic interaction" of coalition advanced military technology and Iraqi ineptitude. But this begs the question, what is the relevance of the RMA to other, more prevalent, forms of conflict besides high-intensity, force-on-force combat? Moreover, how effective would a "tooled-up" RMA-force optimised for high intensity, force-on-force warfare perform against a more skilled enemy than Saddam Hussein's army, particularly one prepared to take advantage of urban terrain, to blend into (and shelter among) the civilian population, to attack through ambush, mines or suicide bombs, and to counter mass and firepower with patience and cunning? In other words, what is the relevance of the RMA to military operations other than war (MOOTW)?

Key assumptions

The answers to this question depend greatly on how one understands the terms RMA and MOOTW. It is therefore necessary to unpack these concepts and make explicit this chapter's understanding of these issues. It is also useful to lay out the underlying assumptions of this chapter, the first being that Clausewitz's definition of war as a "true political instrument, a continuation of political inter-course, carried on with other means" (Clausewitz 1993: 87) is a valid and appropriate starting point for this chapter; that war has simultaneously a military and political character; and, that the balance between the two will vary from one war to another.

> The more powerful and inspiring the motives for war, the more they affect the belligerent nations and the fiercer the tensions that precede the outbreak, the closer will war approach its abstract concept, the more important will be the destruction of the enemy, the more closely will the military aims and the political objects of war coincide, and the more military and less political will war appear to be. On the other hand, the less intense the motives, the less will the military element's natural tendency to violence coincide with political directives. As a result, war will be driven further from its natural course, the political object will be more and more at variance with the aim of ideal war, and the conflict will seem increasingly *political* in character. [Emphasis added]
>
> (Clausewitz 1993: 99)

A second assumption is that wars fought wholeheartedly in defense of a vital national interest, with clearly defined political and military objectives and the commitment of public support coincide with the first image of war in the Clausewitz quote above because their military aims and political objectives are more closely coincident. A third is that the more military and less political a war the more impact the RMA is likely to have. This is because while technology exerts a powerful influence on the military dimension of war its impact on the political dimension is equivocal.

It follows that the impact of the RMA on forms of war other than high-intensity, force-on-force combat (hereafter referred to as conventional war) is incremental, not transformative, in nature. And, while overall the impact of the technologies inherent to the RMA is positive across the spectrum of warfare, the system of systems linking sensor to shooter ever more closely is most relevant, and has greater purchase and therefore the more influence on the outcome, to the rarest forms of contemporary war – conventional – and least relevant to the more common forms – low-intensity, peacekeeping, counter-terror, counterinsurgency and so on.

Understanding military operations other than war

Distinguishing between conventional war and MOOTW

The current term MOOTW is problematic. The tautological definition of the Joint Chiefs of Staff (1995), "operations that encompass the use of military capabilities across the range of military operations other than war," is not especially helpful. Moreover, their list of MOOTW is bewilderingly large – including arms control, combating terrorism, supporting counter-narcotics operations, enforcing sanctions and conducting maritime intercept operations, military support to civil authorities (MSCA), non-combatant evacuation operations (NEO), recovery operations and (paradoxically) supporting insurgencies.

In fact, MOOTW does not seem any more adequate than the old term low-intensity conflict which was itself for many years roundly criticised on the grounds of definitional ambiguity. Both terms represent an omnibus of categories of conflict incorporating a range of variants from asymmetric warfare to guerrilla warfare and many more. On the other hand, it does not advance our understanding of what it is that makes these "unconventional" wars *feel* different. Perhaps it is simply that they tend to be protracted, complex and indecisive – and therefore are best avoided. Lawrence Freedman (2001: 69) has argued just this, saying that the distinction in US policy between war, defined in terms of large-scale combat operations, and operations other than war resides in the post-Vietnam War desire to avoid "Any scenario that threatened another quagmire and had murky purpose ... without a strategy for getting out."

Most analysts, however, tend to conceive of war as a series of "everything-up-to" statements corresponding to at least three levels or types. At the top is nuclear war, a form of war in which the likely level of destruction probably exceeds any conceivable political aim – which may explains why it was never happened. Below that there is total war – everything-up-to the *exchange* of nuclear weapons – a form of war typified by the First and Second World Wars in which the major belligerents mobilised the entirety of their resources in the cause of national survival. Finally there is limited war – everything-up-to total war – where the conflict is constrained in some manner, be that geographically, in terms of the kinds of weapons and or tactics employed, or the political objectives.

It is in this third category that MOOTW is located. What ties them together is not their intensity, which may be high or low – and in any event the perception of intensity is relative to one's closeness to the shooting. Nor is it their size, although one term for this type of war is small war, otherwise how else could it include the Vietnam War, which was high in scale and intensity (indeed, for North Vietnam it was a total war)? It is that they are limited in some respect such that they fall short of total mobilisation. And the most important limitation is not found in the level of military conflict but in its political character, wherein the aim for one or more of the belligerents is neither military conquest nor national survival but the attainment of something which may include the use of military means as only a part of the overall struggle.

Of course, Clausewitzians would observe that *all* wars are limited by politics – that is their defining feature. The difference, therefore, must be one of degree. Total wars, such as the Second World War, tend to have grand but clearly articulated aims such as unconditional surrender. For soldiers, this is relatively straightforward – the annihilation or surrender of the enemy's armed forces and the destruction of his will to resist – which tends to simplify the process of harmonising ends and means. Limited wars rarely, if ever, have such straightforward objectives. As Max Boot (2002: 336–41) has argued: they tend to have neither official declarations of war nor exit strategies; they tend to be fought less than wholeheartedly, especially since they tend to not involve vital national interests and often lack significant popular support; soldiers are as much "social workers;" and they often require multinational command structures.

It is therefore hardly surprising that armies prefer to see their role as being to fight total wars and wish to avoid limited wars. This is perhaps why, although statistically we know that classical warfare between states constitutes only 18 to 20 per cent of wars since 1945, we conceive of it as the norm – conventional warfare – while the myriad small wars and insurgencies which have constituted the actual norm over the last 60 years are called unconventional warfare (Holsti 1996: 22–4). The assumption is that the "proper" business of the military is fighting and winning wars in defence of the nation's vital interests, not messing around in the open-ended, complicated and indecisive business of MOOTW particularly as these rarely seem to involve national interests which might be described by a consensus as vital. This view was evident in a *Foreign Affairs* article by Condoleezza Rice (2000: 53):

> The president must remember that the military is a special instrument. It is lethal, and it is meant to be. It is not a civilian peace force. It is not a political referee. And it is most certainly not designed to build a civilian society. Military force is best used to support clear political goals, whether limited, such as expelling Saddam from Kuwait, or comprehensive, such as demanding the unconditional surrender of Japan and Germany during World War II.

September 11 showed this position to be unsustainable; not even the US could choose its wars; much as it might prefer to fight only what former Secretary of State George Schultz acidly called the "fun" wars (in reference to the then emerging Weinberger–Powell Doctrine), it did not have the luxury. MOOTW are an unavoidable reality. To be sure, the military must be lethal, but it needs to do more than be efficient at killing and breaking (Echevarria 2004: 10). The war on terror and other strategic endeavours of the US and like-minded nations will increasingly require their armed forces to intervene in failed and failing states, which means that they will have to master the skills of nation building, stabilisation, peacekeeping and other missions and equip themselves appropriately.

Nonetheless, until a better one comes into common usage we are stuck with the term MOOTW. However, it is with the understanding that the term is ever more misleading: like it or not, the burgeoning missions which are now called operations other than war will increasingly be seen not as a distraction from the military's "proper" job of preparing for an increasingly unlikely form of conventional conflict conceived in terms of the Second World War but as core military functions (Freedman 1998; Mandelbaum 1998).

Understanding the RMA

The RMA is no less contentious a concept than MOOTW. In fact, there is a range of views on it. Does it really exist? What exactly is it about warfare that the RMA has changed, is changing, or will change in the future? Assuming it is happening, what does that mean in practice for policy planners? One could deliberate at length on the clusters of opinion that have formed in response to these sorts of questions. There are, however, ways in which this discussion of the RMA may be simplified for analytical purposes.

Nearly all conceptions of the RMA hinge on the concept of a system of systems as conceived by Bill Owens (2000), whose vision of the RMA holds that information technology will allow existing weapons systems to function in a more tightly integrated and therefore significantly more effective manner. In technical terms, the system of systems boils down to three key technologies: sensors which collect information about the enemy and the area in which forces will contend; communications which transfer data throughout the force; and precision weapons with which military force can be brought to bear with devastating accuracy. Interestingly, Owens argues that the US has already acquired the system's technological components. What it needs to do in order to take advantage of their power is to network them. This creates a "powerful synergy ... much greater than the sum of the components" leading to the three conditions of combat victory: "dominant battlespace knowledge, near-perfect mission assignment, and immediate/complete battlespace assessment" (Owens 2000: 97–102).

Moreover, while there may be no common definition of the RMA, as the US leads the field in adopting the RMA into its military art, it would seem reasonable to accept the Pentagon's understanding of the concept as the most important. The *Joint Vision* statements attempt to place the precepts of the "system of systems" in an explicit operational framework which it is argued will transform the traditional military functions of manoeuvre, strike, protection, and logistics into entirely new functions called dominant manoeuvre, precision engagement, full dimensional protection, and focused logistics.

The new operational concept to receive the most attention, however, is rapid dominance (Ullman and Wade 1996). What it boils down to is rather than seeking victory in the massing of greater numbers of troops and weapons than one's enemy (i.e., the form of warfare in which fortune favours, in Napoleon's words, the side with the "bigger battalions"), trying to cause "effects" on one's enemy – only one of which might be destruction and then preferably from a range beyond retaliation – in order to debilitate, psychologically overawe and paralyse him without the need for attrition engagements.

RMA capabilities and the challenges of MOOTW

Precision weapons and MOOTW

Leaving aside *long-term* challenges from China or perhaps Russia, the West's military technological superiority over all others is practically unassailable. Only the West is in a position to consider implementing strategies based on high technology that value agility and speed over mass, rely on the ability of sensors to consistently identify the most critical enemy targets in a timely manner, and on weapons precise enough to attack them at long range. The potential payoffs of this are great. Smaller, lighter forces can deploy more rapidly, be sustained more easily and achieve victory more quickly (Freedman 2003: 105–14). What we saw in Iraq in 2003 was an example of how an RMA-ed force could achieve extraordinary battlefield success. It led Vice-President Dick Cheney to claim, "It's been a most impressive performance. Coming on the heels of the Afghanistan operation last year, it's proof positive of the success of our efforts to transform our military to meet the challenges of the 21st Century" (American Forces Information Service, 2003).

Or is it? The first part of Cheney's statement, that the operations to oust the Taliban from Afghanistan in 2001 and Saddam Hussein from Iraq in 2003 were very impressive, may be accepted as fairly unproblematic. The middle part of Cheney's statement, inasmuch as that success in March and April seemed to vindicate the RMA-inspired transformation of the American military, can also be accepted as unproblematic. But the accuracy of its last part, the degree to which it meets future challenges – which is more and more likely to lie in the realm of MOOTW – may be dubious through the prism of the key components of the system of systems: precision, communications and sensors.

Even before either Afghanistan or Iraq, Alan Stephens (1994: 146, 149) had already argued that for the armed forces of the West there will be one basic response to future challenges in low-intensity conflict which will rest on technological superiority – in particular precision-guided munitions (PGM): "fight at a distance" and "where that response cannot be made, combat will be avoided." This is one solution to the problem of MOOTW – just don't do it – the opposite of Nike's famous slogan. One could argue, moreover, that this was the preferred response of the Clinton administration to limited wars, particularly after the death in combat of 18 Army Rangers and Delta Force commandos in Mogadishu, Somalia in October 1993. Subsequent US military actions in Bosnia in August–September 1995, Afghanistan and Sudan in August 1998, Iraq in December 1998, and Kosovo in March–June 1999, all centred on the use of long-range precision strike with minimal or no ground operations. If one's major preoccupation is force protection, precision air power is powerfully alluring, because as Eliot Cohen (1994: 109) has argued, "it appears to offer gratification without commitment."

But appearances can be deceiving. One might rephrase Cohen's statement by stating that air power offers a combat effect (i.e., a target's destruction) without the messiness and danger of contact with the enemy. MOOTW, however, is mostly about the latter and so precision air power is unlikely to deliver much gratification. Furthermore, as Cohen (1994: 120) subsequently says, "Air power can indeed overawe opponents, who know quite well that they cannot hope to match or directly counter American strength. On the other hand, these enemies will find indirect responses." One effective response is to situate critical facilities in civilian or culturally and historically sensitive areas: with sufficient cold-bloodedness, mosques, hospitals, orphanages, museums and the like all make splendid fighting positions.

The problem of hitting only the "bad guys" (assuming you can distinguish them) when they shelter among people whose lives – if hearts and minds are to be won – must not be wasted, is highly acute in MOOTW especially when it takes place in urban terrain. In cities the number of non-combatants and sensitive infrastructure is high, which necessitates restrictive and complicated rules of engagement. Ranges for detection, observation and engagement of the enemy are short. At the same time the battlespace is multidimensional offering attackers many places from which to attack and often the advantage of covered avenues of approach and retreat. If PGM have made tactical and operational manoeuvre suicidal in open terrain, the more desirable will enemy commanders find it to fight in cities because they provide asymmetrical benefits to those who are willing to use the civilian population and infrastructure to their advantage. In other words, on most occasions for most forms of MOOTW, the level of precision required is that of an expert rifleman – an order of magnitude greater than existing stand-off weapons – which is to say that "precision" is still not precise enough. In the words of John Paul Vann speaking of the Vietnam War in 1962,

> This is a political war, and it calls for the utmost discrimination in killing.... The best weapon for killing is a knife, but I'm afraid we can't do it that way. The next best is a rifle. The worst is an airplane, and after that the worst is artillery. You have to know who you are killing.
>
> (Maass 2004)

This leads us to a more fundamental problem with stand-off PGM in MOOTW. Critics would argue that this idea is at the heart of Donald Rumsfeld's vision of "transformation that relies on high-technology weapons systems rather than on soldiers" (Keegan 2003). There is an obvious tension here between the RMA's emphasis on precision and the demand of current operations for manpower. And there are two not-so-obvious tensions. First, as noted above, in MOOTW it is not enough for weapons to be targeted precisely, they must also be employed judiciously and with a fine appreciation of local conditions both literal and political – and this requires more "boots on the ground" than "eyes in the sky." Second and more important, closing the "sensor-shooter gap" presupposes that what war is about is merely identifying targets and shooting them.

Network-centric operations in MOOTW

For conventional combat one can readily see the logical connection from sensors, through command and control to shooters and thereon to "negated objects" and, presumably, mission success. For MOOTW it is not so clear. For one thing, though shooting may prove a necessity it is generally the thing you want your "shooters" to try when all other options have been exhausted. For another thing, success often depends not on negating objects, but on much greater challenges: providing political stability, physical security, psychological reassurance, medical and other humanitarian aid with the ultimate aim of changing people's minds and outlook. A real RMA in MOOTW would expand the arrow in the graph below which links shooters and "negated objects" (Echevarria 2004: 1).

At the heart of the network-centric warfare idea, as indeed of all conceptions of the RMA, is the way in which information technology profoundly magnifies the capability of command, control and communications. The key to success in war is the achievement of information dominance, which boils down to one side knowing more about the other than vice versa in a more timely and exploitable fashion. Doing this involves a number of capabilities: first, the ability to gather relevant information about the enemy's disposition, location and readiness and about the terrain, climate and other factors which may impinge on the combat; second, the ability to process this information – which is to apply some form of judgment to it (not necessarily human); and third, the ability to disseminate this information to relevant commanders in as close as possible to real time. As Sun Tzu (1971: 84) said, "Know the enemy and know yourself, then in a hundred battles you will never be in peril."

Leaving aside for the moment the first point, the ability to gather relevant information, the achievement of information dominance over the opponent in any form of warfare is indubitably a good thing. Yet there is a paradox here too: the network's advantages may inherently involve a trade-off between the tactical and operational levels of war with the strategic. On one level, the problem is the "strategic corporal" phenomenon. As the number of troops on the battlefield gets smaller and the power of their weapons increases the impact that any individual soldier's mistakes may have on the campaign as a whole becomes larger. What makes it especially acute is the telescopic lens effect of the international media, which can make minor blunders, fleeting errors of judgment, or isolated acts of indiscipline into events of a strategic consequence once reserved for general officers. There is, however, a deeper level of the problem. The RMA helps to create an illusion of war that is at odds with its brutal reality and engenders false expectations of its risks and potential costs (McInnes 2002: 136). When things go wrong and these illusions are shattered then support for operations based on them will also suffer; in effect, for commanders on the ground controlling the "optic" of an operation – how it is viewed back home – becomes as important as controlling the action itself.

The trouble is that controlling the "optic" is mission impossible in the Nokia

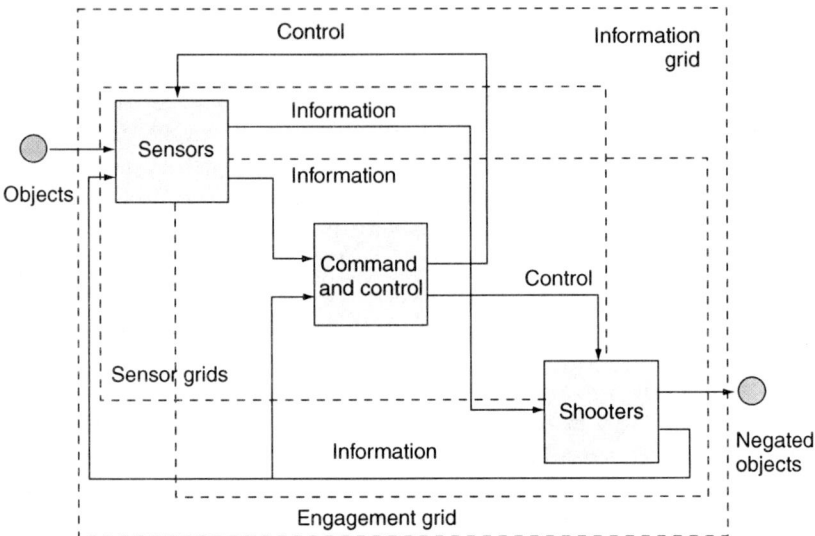

Figure 7.1 Logical model for network-centric warfare (source: Arthur K. Cebrowski and John J. Garstka, 'Network-Centric Warfare: Its Origin and Future', *Naval Institute Proceedings* (January 1998), www.usni.org/Proceedings/Articles98/PROcebrowski.htm).

Age of digital cameras, mobile phones and a ubiquitous Internet. Of course dreadful things have always happened in war (Bourke 1999: 37). The pictures we see now of the abuse of prisoners in Abu Ghraib prison in Iraq are thus not unprecedented. Similar or worse examples can be found in recent history with great ease. What is new is that a networked soldiery that can film anything and store the images on a microchip (or send it on directly via mobile phone) changes the rules of the game. The RMA may be closing the sensor–shooter gap, but it is also closing the gap between what actually happens in war and what the public at home knows about it. As Andrew Marr (2004) noted after the eventually proven false stories about British troops mistreating Iraqi prisoners emerged, "Warfare has depended for centuries on a rampart of silence, a wall of willed incomprehension, between civilians at home and those killing. In a small way, the arrival of digital photographs has broken through that wall."

This is particularly problematic in the context of MOOTW. According to McInnes (2002: 136–7), as a result of the RMA,

> [T]he risks of war and its costs are reduced, minimizing the effect on society at large. The costs are such that society does not suffer in this form of warfare. There is no draft: the individual's engagement is one of choice rather than civic responsibility, and more often than not, that engagement is removed from direct participation in the conflict.

What emerges therefore is a cognitive dissonance in the minds of the public that is presented with contrary messages because it is simultaneously physically disengaged from the action while being able to view the suffering in intimate detail. On the one hand, it may be told that their troops are performing peacekeeping, stabilisation or peace support missions. On the other hand, it is presented with immediate images of graphic violence, death and cruelty. Imagery can be highly deceptive inasmuch as a snapshot says nothing about the frequency or pervasiveness of whatever moment it captures; yet their visceral impact can be intense. What asymmetrical challengers may lack in military power they may make up for with masterful abilities in manipulating outside perceptions by stage-managing events, carefully timing attacks, and feeding information to a gullible, lazy or biased media.

This is particularly a problem in limited wars, as the American military recognises in its *Joint Doctrine for Military Operations Other Than War* (Joint Chief of Staff 1995: II-5), because it impacts directly on two of the unique principles of MOOTW: perseverance and legitimacy. As the doctrine states in respect of the principle of perseverance, "Often, the patient, resolute, and persistent pursuit of national goals and objectives, for as long as necessary to achieve them, is a requirement for success." The more the public experiences confusion surrounding the means and ends of a particular operation the less resolute, patient and persistent it gets. The maintenance of legitimacy, defined as a condition based on the perception by a specific audience of the legality, morality, or rightness of a set of actions, is equally challenging:

> This audience may be the US public, foreign nations, the populations in the area of responsibility/joint operations area (AOR/JOA), or the participating forces. *If an operation is perceived as legitimate, there is a strong impulse to support the action.* If an operation is not perceived as legitimate, the actions may not be supported and may be actively resisted. In MOOTW, *legitimacy is frequently a decisive element.*
>
> (Joint Chiefs of Staff 1995: II-6)

Yet recognition of the problem does not guarantee its resolution. More than anywhere else, the strategic centre of gravity in limited wars seems to reside in the political will of the West to persevere, which, in turn, tends to be determined by the vagaries of public opinion. As a result wars of public relations as much, paradoxically, for the hearts and minds of the domestic audience as for those in theatre have to be waged. What is worrying is how little advantage liberal democracies have in this respect over committed and ruthless but media-savvy opponents. In this sort of information war if the strategic objective (in Rapid Dominance terms) is to control the adversary's will, perceptions, and understanding, then it would seem it is the West which is suffering more greatly from "Shock and Awe." Indeed, the advent of information warfare may yet prove to be a great leveller of military power challenging our notions of military strength. As Bruce Berkowitz (2003: 17) points out,

History will not portray Osama bin Laden as a mere terrorist. Rather instructors at West Point and Annapolis will cite him as one of the first military commanders to use a new kind of combat organization in a successful operation ... technology is driving everyone, terrorist and armies alike, to the same tactics. What is more, most of the technology is commercially available, and thus, it is there for the taking. That is why some of the poorest, most backward countries in the world are able to carry out credible military operations against the richest, most advanced countries, using the same methods.

C4ISR and MOOTW

Thus far we have been talking about *moving* information around within the force as effectively as possible. The third component of the system of systems – sensors – is about *getting* that information in the first place. Knowing more about the enemy than he knows about you is also clearly a good thing. For some RMA proponents, however, it goes further than that: advances in sensor technology, they say, will cause a radical change in warfare. Instead of conducting "-estimate-based" operations using one's *best guess* of where the enemy is and what he might do, technology allows the commander to conduct knowledge-based operations based on *truth*.

> Battlefield sensors are growing in numbers and importance. When coupled with precision-fires capability, sensors can give the friendly force the ability to see and understand the enemy's force and engage it at long range. Given that the requisite munitions and firing platforms are available, an enemy force is threatened with destruction if it is acquired by friendly sensors. As in submarine warfare, the actual destruction of the enemy is in some ways anticlimactic. The real battle is about detection.
>
> (Leonhard 2000: 17)

But then others would disagree; John Keegan (2003) argues that while intelligence is certainly useful in war it is not, by itself, decisive. In fact, even a large information advantage may not guarantee victory because wars are ultimately decided by the skilful and courageous application of sufficient force. Keegan's argument may well be over-stated, but there are a number of problems with the maximalist view of how advances in sensor technology might change warfare. First, as Stephen Biddle (2003: 32) has discovered in looking at the US' recent Afghanistan campaign,

> [T]he war was not purely a standoff affair. Contrary to popular belief, there was plenty of close combat in Afghanistan. Although they were initially taken by surprise, Taliban fighters quickly adapted to American methods and adopted countermeasures that allowed many of them to elude American surveillance and survive US air strikes. These surviving, actively

resisting Taliban had to be overcome by surprisingly traditional close-quarters fighting.

Precision firepower can be extremely lethal. However, if clever, well-trained and disciplined opponents who make consistent use of camouflage and concealment can escape detection in numbers sufficient to mount a significant defense then ground troops will still be required to root them out in the traditional manner. Sensors will of course improve in future; miniaturisation of sensors will make it possible to put them on a wider array of manned and unmanned platforms thus increasing the overall quantity of information. What will improve less slowly is the *quality* of that information since the ability of sensors to see through wood, water, soil, metal and buildings will remain limited by practical engineering restrictions and by the laws of physics.

In short, the way in which an information advantage provided by sensor technology is likely to revolutionise warfare is probably exaggerated. Too much depends on how the information is used – which is not a question of technology per se. Good information can be misinterpreted or lost in the background noise, while bad information can proliferate through the force at light speed. Intelligently employed, information technology can dramatically increase the fighting power of a military force, but it is no substitute for good judgment (Freedman 1998: 52). The exercise of judgement, however, will get more challenging as the overall quantity of information increases. In effect, while one layer of the fog of war may disappear another may descend.

More specifically with respect to MOOTW, there is a grave quality problem the granularity of detail of information which sensors are able to detect is insufficient to what is required for anything like knowledge-based operations to be conducted with confidence. Modern sensors like AWACS, JSTARS, and other reconnaissance assets are of most use for observing conventional orders of battle, particularly in open terrain and especially if it moves. It is much less useful for observing urban militias, rural guerrillas, and "technicals" riding in civilian trucks armed with light weaponry. Still less is it able to determine intent; for example, an individual in civilian clothing firing an AK47 into the air may be attacking low-flying aircraft; on the other hand, he may merely be expressing joy in the local manner. The fact is that one is unlikely to have many useful notions about what is going on in MOOTW without the ability to process information which requires an understanding of local history and culture, which comes from having area experts at hand to advise and from being on the ground for long enough to recognise patterns of behaviour, detect when they change, and act appropriately (Gentry 2002: 96–7).

This is not to say that technology cannot speak to some of the problems of MOOTW. It would seem, however, that the most useful technologies to pursue are those which push the RMA down to the level of the individual infantryman: heightening his ability to behave naturally in close contact with potentially hostile civilians without too great a sacrifice of personal security; helping him distinguish more accurately and quickly combatants from non-combatants; and equipping him with the means to engage them with appropriate levels of force.

A personal weapon which the average soldier could fire with sniper-like accuracy – a few centimetres of dispersion at ranges of several hundred metres – would bring precision down to the level at which it becomes very useful in MOOTW. Light area effect weapons with variable lethality would broaden the range of options available to soldiers in tense but not imminently lethal situations. High-speed, broad-bandwidth communications at the individual soldier level would allow smaller units to practice network-enabled tactics such as "swarming" where those sharing a common picture of the battlespace are able to coordinate their actions without the need for a single commander controlling every move. Advanced lightweight sensors, if linked to relevant databases such as facial features of known enemies, license plate numbers of cars which have been spotted in drive-bys or running road blocks, or merely a reasonably up-to-date record of identity cards would provide soldiers with high quality information, not just high quantity. Light and flexible body armour and uniforms that provide ballistic protection, broad-spectrum camouflage, and protection from weather (hot and cold) would enhance survivability and sustainability. Vehicles with sufficient armour to shelter troops against light arms, mortars and rocket-propelled grenades, yet nimble enough to negotiate narrow streets would increase practical mobility.

One could easily get much more futuristic. Powered exoskeletons could allow soldiers to carry heavier loads, move more quickly, and jump higher. Neural interfaces between man and machine could enhance cognitive power and improve performance of both. "Transdermal nutrient delivery systems" could provide soldiers with all the nutrients and vitamins they require to fight while drugs keep them alert. But this is unnecessary. Much of the more modest technologies above are in fact incorporated, in the American Army's Land Warrior system which includes a new helmet and helmet mounted display, an improved rifle incorporating a multi-function laser, daylight video sight and lightweight thermal sight, improved body armour and load-bearing equipment and a computer with wireless voice and data communications.

Other highly appropriate technologies, particularly for peace support operations, which are now or might soon be developed, include acoustic and infrared sniper detection systems for use in urban scenarios such as Bosnia. Unattended ground sensors (seismic, acoustic, magnetic, optical or infrared) play a useful role where persistent surveillance of vehicular and other traffic in hostile areas or guarding of borders is necessary. Unmanned platforms of all types, particularly cheap and persistent ones, have proven their worth in a range of conflicts. As can non-lethal weapons such as intense light to dazzle, or acoustic weapons to disorient, calmative agents (sleeping gas), sticky foams and super-lubricants in a range of scenarios from crowd control to counter-terrorism (Garwin and Allison 2004). While mine detection/clearance technologies using infrared, x-ray backscatter and nuclear techniques, ground penetrating radar, or chemical "sniffers" for detection and possibly robots or directed energy for safe clearance are badly needed (Feinberg and Maruyama 1998: 107–34).

Conclusion

Thus it would be wrong to suggest that there are not important developments in areas which are quite relevant to MOOTW. Nonetheless, it is also true that none of these technologies plays a central role in any conception of the RMA nor do any of them currently enjoy extraordinary levels of funding. This is unfortunate since technology can provide valuable tools for MOOTW. It is a vital element in a range of areas. Information technology can make training and simulation more accurate and enhance situational awareness. Coupled with advanced sensors and communications it adds to force protection, giving advance warning of dangers and allowing the precise application of firepower. All of this together multiplies effectiveness and increases the likelihood of mission success. But it is also important to recognise what technology cannot do.

For one thing it will not substitute for good political judgement. MOOTW, like war proper, is a political act – in fact according to doctrine "MOOTW are more *sensitive to political considerations*" (Joint Chiefs of Staff 1995: vii). As such it depends on sound strategic decisions being made; as Colin Gray (2002: 270–90) has explained cogently, if you get that wrong then no matter the tactical acumen or technological brilliance of your army you court defeat, because in the long-term good strategy trumps good tactics. Nor can technology make up for poor mission planning and execution, or eliminate the need for appropriate military doctrine, sound training and skilful leadership.

The most expensive weapons and other systems of the American military tend to be focused on doing one thing extremely well: identifying, tracking and destroying enemy targets from ranges beyond easy retaliation. The opportunity costs of this are high because it focuses, and locks in, the vast majority of funding for systems designed for conventional war while neglecting those (including personnel) that are relevant to its more common forms. It is a risky strategy because it is based on a technological hubris that fails to recognise the limitations of the RMA and the peculiar, though potentially strategically significant, ways in which the ongoing changes in warfare work to the disadvantage of the West. And it is a misguided strategy because it plays to a seductive fantasy of the West of fighting only wars that are quick, clean and assured of success when the reality is that the various forms of MOOTW it will face in future will probably be the exact opposite: protracted, dirty and uncertain.

References

Berkowitz, Bruce (2003) *The New Face of War: How War will be Fought in the 21st Century*, New York, NY: The Free Press.

Biddle, Stephen (1996) "Victory Misunderstood: What the Gulf War Tells us About the Future of Conflict," *International Security* 21(2): 139–79.

Biddle, Stephen (2003) "Afghanistan and the Future of Warfare," *Foreign Affairs* 82(2): 31–46.

Boot, Max (2003) *The Savage Wars of Peace*, New York: Basic Books.

Bourke, Joanna (1999) *An Intimate History of Killing*, London: Granita Books.

Clausewitz Carl Von [Michael Howard and Peter Paret (eds and trans)] (1993) *On War*, New York, NY: Alfred A. Knopf.

Cohen, Eliot A. (1994) "The Mystique of US Airpower," *Foreign Affairs* 73(1): 109–24.

Department of Defense (1982) *Field Manual 100–5, Operations*, Washington, DC: Department of the Army.

Echevarria, Antulio J. II (2004) *Toward an American Way of War*, Carlisle: Strategic Studies Institute, US Army War College.

Feinberg, Anthony and Maruyama, Xavier (1998) "Overview of Key Technologies for Peace Operations," in Alex Gliksman (ed.), *Meeting the Challenge of International Peace Operations: Assessing the Contribution of Technology*, Livermore, CA: Centre for Global Security Research, Lawrence Livermore National Laboratory.

Freedman, Lawrence (1998) "The Changing Forms of Military Conflict," *Survival* 40(4): 39–56.

Freedman, Lawrence (2001) "The Third World War," *Survival* 43(4): 61–88.

Freedman, Lawrence (2003) "Prevention, Not Pre-emption," *Washington Quarterly* 26(2): 105–14.

Garwin, Richard L. and Allison, Graham T. (2004) *Nonlethal Weapons and Capabilities*, New York: Council on Foreign Relations.

Gentry, John A. (2002) "Doomed to Fail: America's Blind Faith in Military Technology," *Parameters* 22(4): 88–103.

Gray, Colin S. (2002) *Strategy for Chaos: Revolutions in Military Affairs and the Evidence of History*, London: Frank Cass.

Holsti, Kalevi J. (1996) *The State, War and the State of War*, Cambridge: Cambridge University Press.

Joint Chiefs of Staff (1995) *Joint Doctrine for Military Operations Other Than War*, Joint Pub 3–07.

Joint Chiefs of Staff (1996) *Joint Vision 2010*, Washington, DC: GPO.

Joint Chiefs of Staff (2000) *Joint Vision 2020*, Washington, DC: GPO.

Keegan, John (2003) *Intelligence in War: Knowledge of The enemy from Napoleon to al Qaeda*, New York: Alfred A. Knopf.

Leonhard, Robert (2000) *The Principles of War for the Information Age*, Novato, CA: Presidio.

Kozaryn, Linda D. (2003) "Regime's Collapse Evident, but Hard Fighting Ahead, Cheyney Says," *American Fforces Information Service*, 9 April 2003. Online, available at www.defenselink.mil/news/newsarticle.aspx?id=29138.

McInnes, Colin (2002) *Spectator-Sport War: The West and Contemporary Conflict*, London: Lynne Reiner.

Mandelbaum, Michael (1998) "Is Major War Obsolete?" *Survival* 40(4): 20–38.

Maass, Peter (2004) "Professor Nagl's War," *New York Times Magazine*, 11 January 2004.

Marr, Andrew (2004) "Digital Cameras have Dispelled the Fog of War," *Daily Telegraph* (UK), 12 May 2004.

Owens, Bill (2000) *Lifting the Fog of War*, Baltimore and London: Johns Hopkins University Press.

Perry, William (1991) "Desert Storm and Deterrence," *Foreign Affairs* 70(4): 66–82.

Rice, Condoleezza (2000) "Promoting the National Interest," *Foreign Affairs* 79(1): 45–62.

Stephens, Alan (1994) "The Transformation of Low-Intensity Conflict," *Small Wars and Insurgencies* 5(2): 143–61.

Sun Tzu [S.B. Griffith trans.] (1971) *Art of War*, London: Oxford University Press.

Ullman, Harlan K. and Wade, James P. (1996) *Shock and Awe: Achieving Rapid Dominance*, Washington, DC: National Defense University Press.

8 Small navies and network-centric warfare

Is there a role?[1]

Paul T. Mitchell

Is there a place for small navies in network-centric warfare? Will they be able to make any sort of contribution in multinational naval operations of the future? Or will they be relegated to the sidelines, undertaking the most menial of tasks, encouraged to stay out of the way – or stay at home? If the recent experience of the Canadian navy is any guide, small navies have every right to be concerned about their future in network-centric operations. For while the Canadian navy has achieved a high degree of success within United States (US) naval formations, it has done so only by virtue of highly privileged access. To date, the challenges posed by the revolution in military affairs in general and network-centric warfare in specific have been framed in terms of technology and investment (Gompert *et al.* 1999). The allies and partners of the US are lagging in technology and investment therein, and they need to make significant capital investments in order to catch up. Worse, "dynamic coalitions," developed rapidly to deal with crisis situations, may become the most common form of military cooperation. In such coalitions, detailed, pre-arranged plans and doctrine are likely to be entirely absent. Partners will have had little in-depth operational experience or knowledge of their own capabilities. Technical standardisation will be low; national logistical support may be limited or entirely absent. Significantly, there may be serious questions regarding the professionalism of personnel participating in these coalitions (Spring *et al.* 2000: 5–6).

How dynamic coalitions will function in network-centric warfare is undoubtedly problematic. One commentator has recently suggested that the nature of network-centric warfare may ultimately result in more unilateral (or virtually unilateral) US operations, such as that recently conducted in Afghanistan. In effect, the risk of "clueless coalitions" may drive the US, however unwillingly, toward a more unilateralist military policy, irrespective of that enunciated in its national security strategy (Chekan 2001). The Joint Chiefs of Staff have called for a more "tailored approach to interoperability that accommodates a wide range of needs and capabilities" without implying "access without restraint" (Spring *et al.* 2000: 6). In the unstructured environment implied by the concept of dynamic coalitions, however, the policy restraints upon information sharing, surely the heart of network-centric warfare, may be considerable. As Thomas Barnett (1999: 37) has pointed out: "Not only will our allies have little to contribute to the come-as-you-are party, they won't be able to track the course of the conversation."

This chapter examines the nature of network-centric warfare, the challenges it presents to coalition operations, and some recent developments that seek to overcome these challenges. It uses the Canadian navy's recent and ongoing experience of directly integrating into US carrier battle group operations as a test case. The chapter finds that the principal challenges that will be raised by network-centric warfare are not likely to be technical ones, although undoubtedly these will be formidable. Rather, the most challenging issues for all navies, and small ones in particular, stem from policy. If Canada's example is typical, navies that have less well developed relationships with the US Navy are likely to confront such crippling difficulties in integrating into network-centric warfare-dominated operations as to be excluded from them.

The nature of network-centric warfare

Much of what has been revolutionary in the revolution in military affairs (RMA) is not so revolutionary from a naval perspective (Tritten 1995). Navies have been working with information technology since 1957, when the Canada–United Kingdom–United States (CANUKUS) Naval Data Transmission Working Group, after three years of deliberations, ratified the technical standard for data exchange (Friedman 1997).

Link 11 is more or less standard among Western navies. Primarily used to share tactical information so as to develop what is now known as a "common operational picture" within a task group, Link 11 data is also used by the US Navy to transmit certain engagement orders. However, for many reasons, Link 11 is relatively slow. Because of significant lag times between target detection and the posting of data onto the Link network, its information is not of fire-control quality. Further, it passes to linked ships only the data that has already been processed on board the contributing ship. This occasionally leads to duplicate tracks or conflicting information about the same target. Link 11 demands a high degree of professional competence on the part of track coordinators in order to keep the operating picture "clean" (Friedman 2000).

Network-centric warfare aims at increasing the efficiency of the transfer of maritime information among participating units (or nodes). By optimising the efficiency of operations through information exchange, even small naval formations can generate additional combat power (Smith 2001: 61). Data is manipulated by a series of dynamic and interlinked "grids": sensor grids gather the data, information grids fuse and process it, and engagement grids manage the operations generated (Department of National Defense 2001: 12). Improved operational efficiency results not only from the increased speed at which operations can proceed but also from the "self-synchronisation" that is generated between units (Oxendine 2000: 18). This speed and synchronisation ultimately merge the strategic "recognised maritime picture" with common operational and tactical pictures (Department of National Defense 2001: 11–12). For example, in Canadian ships, the recognised maritime picture is provided to ships by shore-based facilities, whereas ship-based sensors and tactical data links generate local information. At the moment, neither informs the other, which can often lead to discrepancies. With the merging of information into a common pool distributed by linked systems, plans and operations will become much more dynamic. They will be able to react instantly to changes in the battlespace, by virtue of their enhanced awareness of them. For navies having this capability, the result is a competitive advantage, an ability to "lock in success" while locking out enemy initiative (Oxendine 2000: 18).

The original requirement to increase reaction speeds arose in the Cold War in order to deal with hypothesised regiment-sized air attacks on surface ships; the present impetus for speed and synchronisation is the return of fleet operations to their traditional setting, in and around the littorals. The sheer density of maritime and air traffic, the presence of naval, commercial, and recreational maritime vehicles, results in a level of complexity that blue-water operations rarely encounter. This web of activity is made all the worse by the influence of micro-climates, complex oceanography, and unique geographical features. Finally, in the littorals, there are few places where a warship does not stand out, whereas defenders are afforded a multitude of opportunities to hide their forces, whether geographically or through deception, basing them on non-naval platforms (Scott 2002). In effect, naval forces are forced onto an asymmetrical battlefield in the littorals (Pope 2001: 10). In response, however, networked operations permit enhanced speed and synchronisation, which generate predictive planning and pre-emption. Predictive planning and pre-emption in turn provides for three cap-abilities: proactive, "manoeuvrist," effects based operations; integrated force management, allowing synchronisation of missions and resources; and execution of time-critical missions, employing "near optimal weapons pairings."

The most explicit technological development stemming from these concep-tual underpinnings has been "cooperative engagement," which passed its opera-tional evaluation trials in September 2001 (*Defense Daily* 2001b: 1). Cooperative engagement, like Link 11, seeks to develop the common opera-tional picture; unlike Link 11, however, it also aims to co-ordinate threat decisions in real time. Further, it also attempts to distribute fire-control-quality information to participating network nodes (Busch and Grant 2000: 37–9). Cooperative engagement improves a force's ability to share data, even that of a fragmentary nature. For example, because of stealth technology or terrain-masking effects, a ship's sensors may be unable to collect precise and complete information on a particular target. In a formation equipped with cooperative engagement, ships would automatically cue other sensors within the formation, producing a more detailed picture. All this information could then be pooled with the data collected by other more distant ships to assemble a composite picture of the target that no single ship would have been able to generate. Units might thereby receive fire-control-quality information on targets outside their sensor horizons; they could fire weapons before threats appeared to them, allow-ing engagements to take place at maximum distance from the targets (Kerno

1999: 45–7). The end result of all this would be a considerable increase in the time available to make decisions – more time to assess threats and respond – and operations faster than the opponent can sense and respond to himself. Coopera-tive engagement is not the only technical development speeding up the pace and efficiency of naval operations within the US Navy. Much like the private busi-ness world in the last five years, the US military has taken advantage of the Internet to improve the flow of information. The Defence Message System, backed up by the Secret Internet Protocol Routing Network (SIPRNET), has introduced a series of World Wide Web-based applications such as e-mail with attachments, chat rooms, and web pages (Pope 2001: 9–10). SIPRNET in particular seems to have had a revolutionary impact on the planning and conduct of operations within the US military. It has transformed laborious manual proce-dures into rapid electronic ones. This became most evident during Operation Allied Force, when the sheer amount of paperwork forced planners to use elec-tronic formats, "which were substantially easier to create, pass via e-mail, and maintain visibility on." As superiors appended their comments on forwarded messages, it became simpler to track the evolution of commanders' intentions as well (Stuart 2000: 8). Even chat rooms, so ubiquitous among idle teenagers, have a distinctly revolutionary aspect in that they permit the transmission of information (along with attachments of imagery and other intelligence) without radio communication, thus preserving communications security within a theatre (*Defense Daily* 2001a).

Video teleconferencing (VTC) has also led to compressed command and control processes through its ability to span the strategic, operational, and tacti-cal levels. It is particularly useful for staffs that are widely dispersed geographi-cally (Stuart 2000: 7). Video teleconferencing obviates the need to collocate staffs and reduces ambiguity in commanders' intentions (*Defense Daily* 1999). VTC and chat functions collectively permit "distributed collaborative planning," which seeks to assemble problem solvers for rapid and effective response to time-critical situations, while providing access to and ensuring the availability of information resources (Department of National Defence 2001: 20–1). Aircraft carrier battle groups are inherently dynamic given the constant flow through them of ships, personnel, and new technology. It is necessary to control this dynamism rather than be overwhelmed by it; accordingly, a battle group deploy-ment involves a meticulous process of training and planning through which all participating units and individuals become familiar with the synergies between processes, procedures, and systems. The product is a specified battle rhythm, which requires that everything within the group, system, individual, or ship, "not have an adverse effect on communications or information flow." To this end, the battle group proceeds through a series of subunit and unit training exercises. These culminate in the comprehensive task unit exercise that certifies the battle group for basic functions and a final joint task force exercise that combines the carrier battle group with other formations, such as amphibious groups and allied formations (Morua 2000; Peterson 2001: 5).

Operation Allied Force and subsequent operations in Kosovo are widely hailed as beginning the introduction of network-centric operations, and Operation Enduring Freedom in Afghanistan has laid to rest many of the criticisms. This is especially so since that operation saw the confrontation of a high-tech military against a ragtag, guerrilla-type army:

> The Afghanistan operation may ultimately prove to be a boon to the Department of Defence's revolution in military affairs, in which the prize is not territory but information. Only after a clear picture of the battlefield is assured – and that shared with as many weapons platforms as possible – can the maximum potential of PGMs and other high tech weaponry be unleashed both militarily and politically.
>
> (Bender *et al.* 2001)

Particularly impressive has been the manner in which information from a wide variety of sources has been processed and fused for both air and ground forces, thus permitting midcourse updates, engagement zones, "moving target options", and cockpit target imagery (Bender *et al.* 2001: 28).

Equally evident, however, was the initial lack of allied participation in the most secret and demanding operations. While this might have stemmed from a general lack of allied logistical lift, other possibilities must also be considered. As Arthur K. Cebrowski, the godfather of network-centric warfare, has noted, while the US wants its partners to be as interoperable as possible, "not being interoperable means that you are not on the net; so you are not in a position to derive power from the information age" (Howard 2001).

NCW and information barriers

Getting on the net may not be a simple process at all for allies and coalition partners. Essentially, these nations face two distinct challenges: network access may be hampered by technical incompatibilities inherent in their force structures, but it may be obstructed also by design (Pope 2001: 10).

Recent operations in the Balkans have underscored the difficulties of meeting American expectations for rapid, information-dense operations. During Operation Sharp Guard, conducted by the North Atlantic Treaty Organisation (NATO) and the Western European Union in the mid-1990s, the ability of a ship to compile an operational picture was limited at times to its own horizon. Further, the commander of NATO Naval Forces South, in Naples, initially had no timely access to information being collected by units under his command (Germain 1997: 3–4). During Operation Allied Force, "existing data networks were not adequate to support the flow of information of ... data among key nodes of the NATO information grid." Further, the US was unable to pass along "high-fidelity data;" the alliance experienced accordingly difficulties attacking time-sensitive targets, "because of the need for rapid exchange of precision targeting data and continuous precision updates from sensor to shooter until the target is destroyed" (Ladymon 2001: 115).

Although some of these issues later found technical solutions (Operation Sharp Guard units and command centres eventually received old US Navy Joint Operational Tactical System terminals, for example), the "need for speed" in network-centric operations places the whole notion of multinational operations at risk. Interoperability barriers may exclude even close allies. Connectivity problems are the "equivalent of changing to a different railway gauge at each national border" (*International Defence Review* Kiszely 1999); high-tempo operations therefore ultimately become hostages to the units with the slowest information and decision cycles (Smith 2001: 3; Oxendine 2000: 19). Just as pressing,and in the long term even more damaging than technology differentials,may be lack of physical access. Liaison officers have traditionally been exchanged by militaries to ensure the transmission of information among partners, particularly when there are interoperability problems (Scales 1999). Today, liaison officers are often unable to enter US command centres because of security restrictions (Wheatley and Buck 1999: 6). Technology itself may ultimately lead to the electronic equivalents of these physical barriers.

The growing use of video teleconferencing directly raises this issue, because of the classified information frequently involved. In order to access a VTC link, "all users must be on the same level of classification of network and have access to the information on the network" (Pope 2001: 12). The lack of timely written documentation and the instantaneous, experiential nature of VTC hinder any participation by those not on the network (Stuart 2000: 7). As John Kiszely (2002) has pointed out more broadly,

> Full interoperability between forces would depend upon integrated collaborative planning based on the maintenance of a common operating picture and common intelligence inputs. Without appropriate digital communications, this would not be practical, and made all the more unlikely because the US SIPRNET is NOFORN (not releasable to foreign nationals).

Thus, network-centric operations in a coalition or alliance environment may ultimately hinge on rules regarding information release and the ability to exchange information between networks of different security classifications.

The underlying trouble is that the guiding principle of network-centric warfare is to increase the speed and efficiency of operations, whereas coalitions are rarely concerned about combat efficiency. Coalitions are always about *scarcity* – in terms of operational resources, political legitimacy, or both. The trade-off is always in terms of political influence over operational considerations; in coalitions, politics frequently trump efficiency. Neither is policy concerning the release of information oriented around efficiency, but rather security. "Information release and control must be conducted in a manner that prevents damaging foreign disclosure[;] this capability must be demonstrated to information owners" before any transfer can be effected (Spring *et al.* 2000: 7). Information, and what it may imply about the systems that collected it, may be too sensitive to be entrusted to others.

In the absence of clearing houses for information, information disclosure between nations is typically a tedious and cumbersome procedure (Chekan 2001: 9–23; McKerow 2001: 2; Spring *et al.* 2000: 29–34). Further, because the long-term effect of individual disclosures can be difficult to ascertain, and because the career impact of improper disclosure is so serious, "commanders often choose stringent release rules to avoid problems" (Chekan 2001: 11). In this way, concerns about the release of information have dictated separated networks operating at different tempos. As Gary Salisbury, director of command, control, and communications systems for US European Command, characterised the situation in September 2001,

> How do [combined planners] get these national communication and information needs and fit these into a coalition environment? The bottom line is we are generally operating two different networks at two different security levels. We run our networks at a coalition releaseability level that's basically unclassified.
>
> (Kenyon 2001)

It is ultimately these information security policies that prevent allies and partners from operating at the same speed as the American military. Many of the problems of interoperability between allies and coalition partners are the same as those encountered in joint interoperability. Some have suggested that lessons learned from the latter can be applied to coalitions (Black 2000: 5–6). Nevertheless, the intervening variable, not present in joint situations, is that of international politics. The transnational element – particularly as it affects information security – makes coalition and alliance interoperability an order more difficult than joint interoperability.

It would be a gross overstatement to claim that the US is unconcerned by the issue of the release of information. Throughout the 1990s and still today, the US has sponsored Joint Warrior Interoperability Demonstrations (JWIDs), intended to seek technical solutions to common and pressing interoperability problems. These demonstrations have identified several technical solutions; for instance, "Radiant Mercury" and SIREN (Secure Information Release Environment) decision-support software, which speed up the sanitisation and declassification of secret documents (Bender 2001: 45). The 1996 JWID identified the Coalition Wide Area Network (CWAN) as a "golden nugget." CWAN permits establishment of a common operational picture at a "coalition secret" level. Separated (though not entirely) from the SIPRNET by software firewalls and gateways, CWAN was initially introduced in the multinational RIMPAC (rim of the Pacific) exercise series and is currently being widely used elsewhere as well (Spring *et al.* 2000: 17; Pope 2001: 11). Finally, the US Assistant Secretary of Defense for Command and Control has sponsored a series of workshops and seminars among a working group composed of Australia, Canada, Germany, Britain, and the US, with France as an observer. The working group seeks to identify the core needs of information exchange and to establish common doctrine and procedures prior to any operation (Wheatley and Buck 1999: 9).

Dwight D. Eisenhower once famously remarked, "Allied Commands depend on mutual confidence" (Spierto 1999: 3). Like relinquishing command and control, releasing sensitive information is an act of trust between states surpassed only, perhaps, by placing troops under even the limited control of an ally; releasing closely held knowledge places technology, operations, and even personnel at risk (Riscassi 1993). "Trust involves a willingness to be vulnerable and to assume risk. Trust involves some form of dependency" (Chekan 2001: 4).

Thus, we can expect that just as nations have always been unwilling to place complete control of their troops under the control of foreign nations, they will be unwilling to share completely all information they have: "As close as ... Canadian and British allies are in common interests and objectives, there will always be limits to sharing the most highly classified information with these nations" (Pope 2001: 6). In the past, this reluctance did not typically jeopardise operations. However, in network-centric warfare information is the cornerstone of all action; the existence of separate networks operating at different speeds will have an undeniable impact on battle rhythms.

The US is certainly willing to share most of its information with certain partners. For forces of nations not in this privileged club, integration into American networks will be increasingly difficult, depending on how often they operate with the US forces and the degree of trust extended to them. Forces not permitted to take part in planning will ultimately be restricted simply to taking orders – possibly to assume high-casualty or politically distasteful roles (*International Defence Review* 2002). The added risk is that multinational operations will become more and more circumscribed, that allied participation will be accepted only under the most restrictive circumstances. The US is unlikely to hamstring its own military forces or to slow its implementation of network-centric warfare given its obvious benefits. It may decide simply to pass entirely on alliance participation (Carr 1999: 15–16). Policies concerning the release of information would ultimately decide, then, not only the shape and nature of naval coalitions but possibly even their very existence.

Canadian ships in US navy carrier battle groups

One can get a sense of the challenges facing coalition naval network-centric warfare by examining the integration of Canadian warships into US aircraft carrier battle groups. In some respects, this case represents the crucible, for any difficulties faced by Canadians are likely to be considerably more intense for navies outside the bonds of trust that have traditionally connected the Canadian and American navies.

The Canadian navy began arranging to insert its ships into carrier battle groups in the late 1990s in an effort to improve interoperability with the US Navy. Initially, only West Coast ships, operating out of Canadian Forces Base Esquimalt, in British Columbia, were involved. The West Coast fleet had fewer recurring operational commitments (such as the NATO Standing Naval Force Atlantic) than the East Coast command in Halifax, Nova Scotia. Further, the West Coast fleet had a long tradition of operating with the US Navy and was therefore more doctrinally compatible with it than the Halifax squadrons, which had been primarily influenced by their long history of NATO operations.

Since their introduction, the integration of Canadian ships into US carrier battle groups has been an evolutionary process. Canadian ships began as members of the Maritime Interdiction Force in the Persian Gulf, later gradually moving into actual battle groups as mutual familiarity improved. What started first as an operational initiative eventually gained an explicit strategic stature (in the Canadian context), when it became Department of National Defence policy to improve interoperability with its allies, particularly the US. The department now seeks to develop and maintain "tactically self-sufficient units," capable of substantial military contributions while asserting their Canadian identity. A ground-forces equivalent would be the role Canadian Coyote LAV IIIs, armoured reconnaissance vehicles, played in Bosnia, Kosovo, and now Afghanistan. Dan McNeil, Director for Force Planning and Program Co-ordination, has recently remarked:

> We will never be able to field strategic level forces.... We're not ever going to be in that game. We're going to be fielding tactical units. [However,] if you properly use tactical units, you can achieve strategic effect. That is what we are trying to do.
>
> (Hobson 2001)

A revolutionary aspect of these carrier battle group operations has been the fact that individual Canadian ships have often replaced US ones. This arrangement has been of mutual benefit; the US has been able to address its shortages of frigates and destroyers, and Canada has been afforded professional opportunities that it could not hope to obtain on its own. These opportunities include not only extended operations in groups larger than those the Canadian navy typically sends to sea, but also exposure to assets not in the Canadian order of battle – carriers, cruisers, and nuclear submarines.

Canada has thus become a member of a select club, enjoying special access to the command and control concepts developed by the US Navy as it travels down the road of network-centric warfare, as well as to military support not normally offered to allies. Finally, carrier battle groups operations enable the Canadian navy to develop professional skills in the areas of littoral and interdiction operations, for which there is no opportunity in North American waters.

At the same time, such deployments stress the mutual dependencies and vulnerabilities that are central to every good coalition operation. For the Canadian navy, given the relative scarcity of Canadian ships (Canada has only 12 *Halifax*-class frigates), each unit deployed has value out of proportion to its ultimate contribution to a carrier battle group. Obviously, sending such ships into the Persian and Arabian Gulfs, as is typical, is far more dangerous than assigning them to the standard fisheries patrols in Canadian waters they would most likely be conducting otherwise. Similarly, by replacing a US ship with a Canadian one, rather than simply augmenting the group, the US Navy is placing considerable trust in the professionalism and competence of Canadian crews; as one battle group commander has declared, "We need to be ready to go on game day – and when we play, every game is game day" (Peterson 2001: 7). Accepting a Canadian ship into a battle group also constitutes a commitment to look after that ship.

To ensure that they are not liabilities for their new battle groups, Canadian ships participate in the same exercises and workups that all American ships do. Similarly, they carry the latest revisions of the Global Command and Control System–Maritime (GCCS–M) and conduct training to ensure that they can share and use the information and imagery distributed on that system. The Canadian navy has been increasingly challenged by such upgrades, however, due to the legacy systems on board its ships. The CCS330 system that controls the ship displays in the operations rooms of the *Halifax* frigates and *Iroquois*-class destroyers is a closed-architecture system based on a unique operating system and military-specific software and hardware. State of the art ten years ago, it is increasingly becoming a maintenance problem and, even more seriously, has a very limited capacity for integration with new systems. New capabilities, like GCCS–M, must be added to Canadian ships on a stand-alone basis. Canadian display terminals, as a result, cannot send or receive operational messages; tactical networking requires separate consoles; and the information provided by systems like GCCS–M and the Canadian equivalent of the SIPRNET, known as MCOIN III, become effectively "stove-piped." The result is a cluttered operations room where decision makers must consult a number of systems in order to gather all the information necessary to perform their jobs – obviously not the most efficient arrangement in the heat of battle (Department of Defence 2001: 17).

Interestingly, the Canadian navy's effort to remain abreast of the fast-moving electronics revolution in command and control technologies is not being driven by US requirements. The US is pleased that Canada strives to prevent gaps in capabilities. However, Canadian naval officers stress, it is the long history of naval cooperation and overall familiarity between the navies that has facilitated these exchanges, not the technical "kit" installed aboard Canadian ships. The difficulties Canadian ships typically encounter in integrating themselves into US battle groups largely arise from the issue of accessibility.

In carrier battle group operations, as noted, the Coalition Wide Area Network is the principal means for coordinating action between Canadian and American ships; the US Navy is gradually migrating its command, control, communications, planning, and execution functions to web and other digitally based delivery methods, notably the SIPRNET. However, CWAN and SIPRNET have mutual interface limitations. E-mail can pass between the two systems as long as the US user has a CWAN account. Nevertheless, a security firewall strips off attachments before admitting messages into the CWAN. Thus a Canadian recipient may receive a commander's directive but not the supporting and amplifying information that originally accompanied it. Furthermore, messages from SIPRNET users without registered CWAN accounts will not reach Canadian ships, which may thereby miss important items.

The growing use of chat features to plan and coordinate has also been noted, and CWAN has such features. However, there is no interconnection between SIPRNET chat and CWAN chat. In order for a Canadian ship to participate in a session with US counterparts, a CWAN liaison officer must type into CWAN what was entered onto the SIPRNET system. Any attachment must be "air-gapped" onto CWAN, which can be quite a complicated procedure, involving multiple transfers between networks (SIPRNET to NATO Information Tactical Display System to MCOIN III). As there is frequently only a single Canadian liaison officer on the carrier, accordingly, transfers between the two systems are likely to be delayed when that officer is not on watch. Canada urges the US flag-ships to man the CWAN terminal during these times, but it is likely to be over-looked in periods of high operational tempo – just when the Canadian ships most need the information.

Finally, the web features of SIPRNET are limited on the CWAN side. CWAN supports web pages, but they contain only information placed there by coalition partners. In a US-run operation, the majority of the information needed will be originating from the US. There is no direct connection between SIPRNET web pages and CWAN web pages; web files must be "air-gapped." As a result, CWAN and MCOIN III are often out of date, sometimes by days. Furthermore, CWAN information is likely to be only a snapshot of that available to SIPRNET, without the functional links that it has on the US side, limiting the ability of coalition officers to surf for more information. Finally, the carrier is usually the only US ship in a battle group with a CWAN terminal, in which case it is the sole unit capable of posting information there – making it all the more possible that important information will not be posted at all.

Trust and unilateralism

There may be nothing available but inefficient, work-around solutions to these problems. The real difficulty is not so much technical as policy oriented. The natural desire to protect sensitive information is at the root of all these issues, and it is not unique to the US – MCOIN III is a Canada-only system, just as SIPRNET is US-only. We should not expect this sensitivity to disappear any time soon; in fact, 11 September 2001 doubtless heightened it. Software that facilitates the release of information helps to move information onto coalition networks in a timely fashion, but they are not gateways to the information that US Navy officers use on a day-to-day basis. This results in two quandaries for Canadian ships. First, they often operate without even basic operational-procedure manuals; some publications have not been classified as releasable to Canada or to the CWAN. Without such formal guidance, US Navy officers are generally reluctant to release even what is seemingly innocuous data for fear of making mistakes that could have repercussions for their careers (O'Brien 2001: 6). Second, since the makeup of a carrier battle group is not permanent, information-sharing protocols must be re-brokered for each deployment. Sometimes gaining access is a question of proving one's bona fides to the battle group; sometimes the battle group staff is simply unaware what information has been passed, or is otherwise available, to the Canadian ship. Often such problems are resolved when the battle group commander becomes aware of them, but the necessity to approach "the flag" for such matters highlights the impediments to network operations in a coalition environment.

The Canadian experience with US carrier battle groups is instructive in both positive and negative senses for the overall question of network-centric operations in a coalition environment. It is positive in demonstrating that despite technical limitations and differences between two navies, effective cooperation can be achieved in the modern naval environment. Once willingness to cooperate and a basis of trust between two forces has been established, technology is not an impassable barrier. Canada's close experience with the US may be helpful to other navies. In its vision document *Leadmark* (2001: 107), the Canadian navy has proposed to develop a "Gateway C4ISR" function that would allow less capable navies to integrate themselves into network-centric operations. The Canadian navy has performed such a function in the past. During the Gulf War, among the deciding factors in the selection of Canada to lead the Combat Logistics Force were its excellent interoperability with the US (a proposed French ship, *Doudart de Lagrée*, "lacked good communications interoperability"), its multinational crews, and its remaining legacy communications systems (with which Canadian ships could talk with more or less all warships present) (Morin and Gimblett 1997: 181–2; Miller and Hobson 1995: 156). At present, Canadian ships play an important intermediary role in passing on information to other coalition partners in the Arabian Gulf.

However, there is a very large caveat – the relationship between the Canadian navy and the US Navy took decades to evolve, and even so significant impediments remain to the seamless integration of forces that network-centric warfare demands. Further, while carrier battle groups must be prepared for all warfare eventualities, Canadian ships have participated predominantly in maritime interdiction. One wonders how welcome even Canadian ships might be in an operation dominated by strike warfare, against an asymmetric surface threat, in the littorals. Finally, the security demands of US military networks are likely to be troublesome indeed for navies without the privileged access afforded to Canadian ships and crews on the basis of long-shared operational experience and a wealth of trust. Indeed, if the Canadian experience indicates that coalition network-centric operations are possible, it also indicates that the price of admission will remain very high. In a dynamic coalition environment, professional trust will be critical, and the height of the bar will be set by both technology and policy. Because of the crippling effect of slower networks or non-networked ships in such a setting, issues pertaining to the release of information may be a stimulus to American unilateralism.

Note

1 This chapter first appeared in the *Naval War College Review*, Spring 2003, Vol. LVI, No. 2.

References

Barnett, Thomas B. (1999) "The Seven Deadly Sins of Network Centric Warfare," *U.S. Naval Institute Proceedings*, January.

Bender, Bryan (2001) "JWID Puts Information Sharing System to the Test," *Jane's Defence Weekly*, 16 August.

Bender, Bryan, Berger, Kim, and Koch, Andrew (2001) "Afghanistan's First Lessons," *Jane's Defence Weekly*, 19 December.

Black, Michael B. (2000) *Coalition Command, Control, Communications, Computer and Intelligence Systems Interoperability: A Necessity or Wishful Thinking?* Kansas: Fort Leavenworth.

Busch, Daniel and Grant, Conrad J. (2000) "Changing the Face of War: The Co-operative Engagement Capability," *Sea Power*, March: 37–9.

Carr, James (1999) *Network Centric Coalitions: Pull, Pass, or Plug-in?* Course paper, Newport, RI: Naval War College.

Chekan, Robert (2001) "The Future of Warfare: Clueless Coalitions?" Unpublished paper, Toronto: Canadian Forces College.

Defense Daily (1999) "Defense Watch," 18 October.

Defense Daily (2001a) "Center Outlives Network Centric Warfare Concepts Challenges," 23 March.

Defense Daily (2001b) "CEC Passes Through Successful OPEVAL, Navy Says," 25 September.

Department of National Defence (2001) *The Canadian Navy's Command and Control Blueprint to 2010*, Ottawa: National Defence Headquarters.

Friedman, Norman (1997) *World Naval Weapons Systems 1997–1998*, Annapolis, MD: Naval Institute Press.

Friedman, Norman (2000) "CEC and Fleet Defence," *RUSI Journal*, 145(5): 30–6.

Germain, Eric Francis (1997) "The Coming Revolution in NATO Maritime Command and Control," *MITRE Technical Papers*, Arlington, VA: MITRE Corporation.

Gompert, David C., Kugler, Richard L., and Libicki, Martin C. (1999) *Mind the Gap: Promoting a Transatlantic Revolution in Military Affairs*, Washington, DC: National Defence University Press.

Hobson, Sharon (2001) "Canada Aims for Defence Interoperability with the U.S." *Jane's International Defence Review*, 1 January: 54.

Howard, Peter (2001) "The USN's Designer of Concepts," *Jane's Defence Weekly*, 3 October: 29.

International Defense Review (2002) "General Warns over Digitisation Split," 1 January.

Kenyon, Henry S. (2001) "Alliance Forces Move toward Unified Data Infrastructure," *Signal*, September: 21.

Kerno, Robert (1999) "Co-operative Engagement Capability and the Interoperability Challenge," *Sea Power*, March: 45–7.

Kiszely, John (1999) "Achieving High Tempo: New Challenges," *RUSI Journal*, 144(6): 47–53.

Kiszely, John (2002) "General Warns over Digitisation Split," *International Defence Review*, 1 January.

Ladymon, Joseph M. (2001) "Network Centric Warfare and Its Function in the Realm of Interoperability," *Acquisition Review Quarterly*, Spring/Summer: 111–20.

Leadmark: The Navy's Strategy for 2020 (2001) Ottawa: Department of National Defence.

McKerow, Gary (2001) "Multilevel Security Networks: An Explanation of the Problem," *SANS Information Security Reading Room*, 5 February. Online, available at www.sans.org/reading_room/whitepapers/standards/.

Miller, D. and Hobson S. (1995) *The Persian Excursion: The Canadian Navy in the Gulf War*, Clementsport, NS: Canadian Peacekeeping Press.

Morin, Jean and Gimblett, R. (1997) *Operation Friction*, Toronto: Dundurn Press.

Morua, Michael L. (2000) "The Carrier Battle Group Force: An Operator's Perspective," paper delivered at Engineering the Total Ship (ETS) 2000 Symposium, Gaithersburg, MD, 21–3 March.

O'Brien, Kevin (2001) "Europe Weighs Up Intelligence Options," *Jane's Intelligence Review*, 1 March: 6.

Oxendine IV, Elias (2000) "Managing Knowledge in the Battle Group Theatre Transition Process," thesis, U.S. Naval Postgraduate School, Monterey, CA, September.

Peterson, Gordon I. (2001) "Ready to Go on Game Day: At Sea with the USS *Theodore Roosevelt* Battle Group," *Sea Power*, September: 5.

Pope, William R. (2001) "U.S. and Coalition Command and Control Interoperability for the Future," thesis, US Army War College, Carlisle, PA, April.

Riscassi, Robert W. (1993) "Principles for Coalition Warfare," *Joint Forces Quarterly*, 60(16): 58–71.

Scales, R.H. (1999) "Trust, Not Technology Sustains Coalitions," in Robert H. Scales, Williamson Murray, Paul K. Van Riper, and John A. Parmentola (eds), *Future Warfare*, Carlisle, PA: US Army War College.

Scott, Richard (2002) "Survival of the Fittest," *Jane's Defence Weekly*, 23 January.

Smith, Edward (2001) "Network-centric Warfare: What's the Point?" *Naval War College Review*, 54(1): 59–75.

Spierto, Thomas (1999) "Compromising the Principles of War: Technological Advancements Impact Multinational Military Operations," course paper, Newport, RI: Naval War College, 5 February.

Spring, S.C., Gormley, D.M., McMahon, K.S., Smith K. and Hobbs, D. (2000) "Information Sharing for Dynamic Coalitions," *VPSR Report 2836*, Arlington, Va: Pacific Sierra Research, December.

Stuart, Robert M. (2000) "Network Centric Warfare in Operation Allied Force: Future Promise or Future Peril?" course paper, Newport RI: Department of Joint Military Operations, Naval War College, 16 May.

Tritten, James (1995) "Revolutions in Military Affairs: Paradigm Shifts and Doctrine," *A Doctrine Reader*, Newport Paper 9, Newport, RI: Naval War College Press.

Wheatley, Gary and Buck, D. (1999) "Multinational Command and Control: Beyond NATO," paper presented to 1999 Command and Control Research and Technology Symposium, Naval War College, Newport, RI.

Part III

Impediments to transformation

9 Organizational culture and change

The revolution in military affairs, counterinsurgency, and the US Army

Elizabeth Kier

Even among its ardent proponents, there is a faint but recurring acknowledgment that the Revolution in Military Affairs (RMA) means more than sophisticated information technology and new ways of warfare. "We must transform not only the capabilities at our disposal," former Secretary of Defense Rumsfeld explained, "but also the way we think" (Department of Defense 2003: 1). Yet despite all the references to cultural change (for instance Cebrowski and Garstka 1998; Krepinevich 1999/2000), the RMA debate has largely focused elsewhere. This chapter addresses this lacuna by drawing on examples of the potential relationship between the United States (US) Army's culture and its transformation.

This chapter first outlines the theories of organizational culture. It then addresses the question of how the RMA might relate to a military's culture. It suggests that while the RMA accords well with most aspects of US Army culture relevant to conventional operations, the outlook is less promising for counterinsurgencies (COIN). Given that asymmetrical conflicts are the most likely threat to US security, this conclusion is troubling: the US Army may be increasingly less capable of accomplishing one of its most likely missions. The chapter then examines how cultural change in organizations occurs, and concludes with cautionary advice about the limits of engineering cultural change in the military.

Organizational culture

The literature on organizational culture is as diverse as culture itself. Theorists disagree about how to define it, what it is, and how best to study it (Alvesson 2003: Martin 2002). Schein (1992: 111) defines it as

> A pattern of shared basic assumptions that [a] group learned as it solved its problems of external adaptation and internal integration, that has worked well enough to be considered valid and therefore, to be taught to new members as the correct way to perceive, think, and feel in relations to those problems.

Organizational culture is a set of shared meanings or taken-for-granted assumptions.

Theorists classify the literature on organizational culture as falling into two broad camps (for instance, Smircich 1983: 339–58; also Martin 1992; Schultz and Hatch 1996: 529–57; Barley *et al.* 1988: 24–60). Practitioners and management consultants tend to treat organizational culture as a variable: it is something that an organization *has*. Executives can manipulate it to build employee commitment to organizational goals, and the stronger (or more cohesive) the culture, the better the performance (Deal and Kennedy 1982; Peters and Waterman 1982). In contrast, organizational theorists adopt a more anthropological stance: organizational culture is something an organization *is*. It emerges from social interaction. Here, organizational culture is more a cognitive or symbolic system that influences how members perceive their world than the means of generating affective ties to the organization. Whether the culture enhances organizational performance has more to do with whether its beliefs suit the organization's environment than whether the culture is "strong" or widely shared in the organization.

This view of organizational culture as a cognitive or symbolic system implies that it is both a compass and a blinder (Lim 1995: 16–22; Wilderom *et al.* 2000). Culture makes organizational life possible: complex interaction would be unimaginable if specific actions did not call for specific responses and if members had no understanding of the effects of their actions on others. Some predictability is necessary to organizational effectiveness. But organizational culture also narrows one's view: it makes some things possible, but others unimaginable. It shapes the range of possible reactions to problems and to what is – or is not – defined as a problem. For example, for more than 60 years, the US strategic nuclear community's "organizational frame" led them to ignore the devastating and predictable fire effects of nuclear weapons, focusing instead on predicting nuclear blast effects, and thus underestimating the destructive power of nuclear weapons (Eden 2003).

Before discussing what this view of organizational culture reveals about how RMA developments will influence the US Army's ability to wage COIN, it is useful to first clarify two ways *not* to use organizational culture. First, actors often use references to organizational culture to avoid individual accountability: they often point to "cultural" problems when confronting evidence of organizational failure. Reactions to the intelligence failures prior to 9–11 illustrate this tendency. Despite evidence of individual failures, explanations often focus on cultural problems in the organizations. When criticized for not disciplining any individuals, an FBI official explained that individuals could not be held accountable because the problems were "systemic" (Jehl and Lichtblau 2004). Since culture is an attribute of an organization, and not an individual, it is an attractive excuse for actors hoping to skirt responsibility.

Second, organizational culture is not simply a useful tool to explain irrational or dysfunctional outcomes. Analysts often point to aspects of an organization's culture that constrain its effectiveness, but culture is as much about explaining good outcomes as bad ones. To imply otherwise is to assume that rational actors inhabit a culture-free world, while irrational ones are trapped in collective understandings; the ubiquity of culture makes this belief implausible. Instead, the concept of organizational culture is best viewed as providing explanations for reasonable behavior, which is behavior that makes sense given the shared meanings within that organization. For example, in his study of wartime cooperation, Jeffrey Legro (1995) uses organizational culture to explain why some restraints are respected in warfare, regardless of whether these restraints are militarily optimal.

A military's culture can contribute to organizational performance. For example, given its culture – such as its ready embrace of technology and its focus on the use of decisive and overwhelming force – the RMA is likely to enhance the US Army's combat effectiveness in major theater wars. The German Army's tradition of "mission-oriented command," allied to its more "professional" or "enabling" structure that encouraged subordinates to take the initiative to accomplish the designated mission, is another example of how particular values or beliefs can enhance combat performance (Soeters 2000: 468–70; Soeters and Recht 1998). This attitude helped pave the way for the German Army's adoption of the blitzkrieg doctrine and it may also be well-suited to the demands of networked warfare. The Israeli Army has a similar command philosophy so its culture may also help it exploit the combat potential of flattened organizations with decentralized command (Demchak 1996: 187–8). As the next section explains, military culture can lower combat effectiveness, but that is not always the case. Culture can lead to both insights and blindspots.

US Army culture

Because the US Army dominates large-scale conventional conflicts, it is especially important to consider how the RMA will influence the Army's ability to meet unconventional threats. Few adversaries would choose to confront the US Army on its preferred turf so, like it or not, asymmetrical conflicts are the most likely challenge the Army will face (Catanzaro 2001; Krepinevich 1999/2000). However, the US Army is not well prepared to meet these threats. Two important aspects of its culture – its preference for large-scale conflict and its belief that "one size fits all" – discourage it from developing expertise and capabilities tailored to the lower end of the military spectrum.

US Army culture and unconventional wars

The US Army's *preference for conventional warfare* and its corresponding disdain for "constabulary duties" is well known (Avant 1996–97: 51–90; Campbell 1998; Cassidy 2004a: 73; Cassidy 2004b: 83–4, 93, 99–100; Krepinevich 1986). The Powell Doctrine is the most visible manifestation of this attitude. As the conventional wisdom has it, "we don't do windows, jungles, cities or guerillas" (Krepinevich 2003: 2).

This preference blinds the Army to the risks of low-intensity conflicts. Encouraged by the Bush administration's wildly optimistic expectations for stability in post-war Iraq, this seems to have blinded Army commanders from recognizing the need to prepare for stability and COIN operations in Iraq, despite contrary warnings from the intelligence community that Hussein's regime might be preparing an insurgency (Ricks 2004). It was only in the spring of 2004, over a year after the initial invasion, that US Army commanders began to see the threat they faced. The reconstruction of the Iraqi military first focused on building conventional forces designed to defend Iraq's border (as opposed to specialist COIN units) (Cordesman 2004: 1–15, 23–6). A US Army historian explained: "A reluctance in even defining the situation ... is perhaps the most telling indicator of a collective cognitive dissidence on part of the US Army to recognize a war of rebellion, a people's war, even when they are fighting it" (Ricks 2004). Given the raging insurgency in Iraq, what is perhaps most revealing are the assumptions in a November 2004 war-game simulating an American-led regime-change operation in Iran. The game was based on current Pentagon thinking. It simply repeated the same mistakes made in Iraq (Fallows 2004).

An important corollary to the US Army's preference for conventional warfare is the "one size fits all" belief – that highly-developed conventional capabilities will be sufficient for the full range of military operations (Krepinevich 1986: 35; Krepinevich 2004b: 51, 60). The RMA reinforces this illusion. Instead of developing constabulary forces to respond to unconventional threats, the RMA apparently reinforces this "one size fits all" belief. Integration of RMA technology will produce, its advocates argue, a military able to adapt to any foreseeable contingency. The Joint Chiefs of Staff's *Joint Vision 2010* (1996: 2, 25–6) calls this "Full Spectrum Dominance." Notwithstanding these rhetorical nods to unconventional warfare, the Department of Defense (DoD) has developed RMA capabilities almost exclusively for major theater war. It has not meaningfully explored the applicability of the RMA to unconventional warfare or whether this full spectrum dominance will be realized (Grant 2000: 1; Krepinevich 2004b: 52, 60, 103–8).

US Army culture, RMA, and COIN

Although the US Army could use aspects of the RMA, such as precision weaponry, to better fight COIN, this chapter argues that the RMA is more likely to reinforce Army biases and further divert it from the preparation necessary to meet an insurgent threat. The RMA is likely to further limit the US Army's COIN ability when combined with four aspects of the Army's cultural beliefs: overwhelming force, casualty aversion, faith in technology, and a un-Clausewitzian conviction in a sharp divide between the military and the political. In the context of the US Army's culture, the promise of the RMA and the demands of counterinsurgency are orthogonal. Before exploring how these four aspects of US Army culture likely influence the Army's use of the RMA during a COIN campaign, this chapter first reviews three key COIN principles.

While the key to success in conventional warfare is destroying an adversary's military assets, COIN principles are different (Krepinevich 2004a). The focus is on gaining the support of the wider population or at a minimum providing security to populous areas. From the basic idea that the people are the foundation of the enemy's strength, flow three principles about how to wage a counterinsurgency. First, and most important, a credible campaign focuses on cutting the enemy's popular support (or their infrastructure), not destroying the enemy's forces. Denied popular support and unable to use fear to incite compliance, enemy forces will lose their cover, their intelligence on government forces, as well as access to food, medicine, and new recruits. The guerrillas would become, to paraphrase Mao, "a fish out of water."

The second observation about COIN is the importance of firepower restraint, not its widespread application. The perception of needless civilian casualties will only feed the insurgent's popular support. For example, the brutality of the British-hired German mercenaries in the US War of Independence helped swing undecided colonists to the revolutionary cause (Singer 2003: 33). This principle follows from the final principle for fighting an insurgency – that is, the recognition that the conflict is not primarily a military one. Population security is the first priority, and resources must also be devoted to public works and socio-economic reform if the people are to be persuaded that their future lies with government forces. Historians credit General Sir Gerald Templer for turning the tide in Britain's favor during the Malaya Emergency. Templer understood that the campaign against the guerrillas required more than military actions:

> [A]ny idea that the business of normal civil government ... [is] separate must be killed for good and all. The two activities are completely and utterly interrelated ... The shooting side of this business is only 25 percent of the trouble and the other 75 percent lies in getting the people of this country behind us.
>
> (Smith 2001)

The RMA can in theory benefit COIN. The fewer the civilian deaths, the fewer the risks of alienating the civilian population that government forces must win over. However, a key feature of the RMA is also lethality. This focus on destruction, albeit with precision munitions, may cause more problems than it solves, as the promise of pinpoint accuracy may distract the US Army from its primary objective of denying the insurgents' access to the wider population. RMA capabilities may render unattractive otherwise necessary foot patrols of population centers. In the face of the rising insurgency in Iraq, the US military has increasingly relied on airpower to target insurgent hideouts. This development worries one defense analyst: "Americans seem to believe that airstrikes will wear down the insurgency and buy time for US training of Iraqi security forces. But you have to wonder whether we're radicalizing the Iraqi civilian population in the meantime" (Burns 2004). Even with correct intelligence and flawless accuracy, airstrikes increase the possibility of civilian casualties that may fuel support for the insurgency, regardless of whether claims of civilian deaths are substantiated. Offensives designed to destroy enemy forces while leaving populations unsecured play into the insurgents' hands: they can continue to replenish their forces and draw attention to civilian casualties.

The RMA's focus on lethality is likely to reinforce US Army beliefs about the use of force, particularly about the application of *decisive and overwhelming firepower*. Colin Powell explained, "once a decision for military actions has been made, half measures ... extract a severe price in the form of a protracted conflict which can cause needless waste of human life" (Cassidy 2004b: 110). The US experience in Somalia illustrates this tendency. Units on patrol were told to employ maximum controlled violence and the US military's indiscriminate use of firepower likely led to many civilian casualties. As one US general quipped: "if we had put one more ounce of lead into South Mogadishu ... I believe it would have sunk" (Thornton 2004: 83–106).

The US military's limited preparation to use non-lethal weapons (NLW) in Iraq, despite their successful use during the withdrawal of United Nations (UN) peacekeepers from Somalia and their obvious application to stability operations, also illustrates the US Army's attitude about the use of force. The Army deployed some NLW to Iraq, but it distributed them to military police units, not combat units (Allison and Kelley 2004: 8). NLW do not fit with assumptions about the use of overwhelming force. Similarly, in an ironic perversion of Mao's dictum to be a "fish in the sea of the people," US forces have an acronym for one of their tactics in Iraq: FISH or Fighting in Someone's House, which calls for throwing a hand grenade into each room before checking it for "unfriendlies" (*The Economist* 2005). These and similar heavy-handed tactics may save American lives in the short term, but are likely to be counterproductive over time because they fail to place a premium on population protection. As the former British Foreign Secretary Robin Cook warned,

> What we have done in Fallujah, what we have done repeatedly across Iraq, is we have treated a whole community as if it is the problem. The result ... is that the whole community has responded by saying very well, if we are being treated like them we ourselves will be sympathetic and give support to them.
>
> (Lyons 2004).

Not all militaries share the American belief in overwhelming force. The British military, for example, emphasizes minimum force (Mockaitis 1990; Thornton 2004; Williams 1998). Central to this British concept is the principle of restriction: that force be turned to only with extreme reluctance. In unconventional conflicts, the key British objective is to deter and control violence, not to use decisive force to destroy the adversary. Operations in Iraq provide numerous examples of this Anglo–American divide over the use of force. The commanding officer of British forces in Basra explained: "if there's a man on a corner with an RPG, I can shoot him and someone else will just take his place" (Burke

2003; Rudebeck 2003). In contrast, bumper stickers on US vehicles state: "Keep 50 meters away or deadly force will be applied," and US forces say that they both shoot at any Iraqi handling a phone near a bomb-blast and fire on cars getting within 20–30 meters. As one lieutenant explained: "it's kind of a shame, because it means we've killed a lot of innocent people" (*The Economist* 2005). It also creates great military friction between British and American soldiers: "We must be able to fight with Americans [but] that doesn't mean we fight as Americans" (Evans 2004; Morris 2004).

Some of these differences are attributable to the varying levels of threat that British and US forces face, but the contrast still holds when personnel of both armies are in similar situations. During fighting in the southern city of Umm Qasr during the initial combat operations, a US Army company fired missiles and mortars for four hours at a group of buildings that they suspected held Iraqi forces. The following day, shots were fired at British forces from a house in the area. Six British soldiers dismounted their vehicles, walked to the building, and captured two gunmen; the operation lasted ten minutes with no collateral damage. A British officer commented on his ally's tactics: "For the Americans, there just does not seem to be anything between peace and all-out combat. Their military doctrine remains one based on the use of overwhelming firepower in every circumstance" (*The Economist* 2003). The contrast is also evident in British operations in Basra and Amara, where British forces faced the heaviest fighting since the Korean War. British forces defended their location but did not launch a counteroffensive. This was unlike the US assault in Najaf, which left the old city in ruins.

These anecdotes echo contemporary and historical studies that distinguish British and American army attitudes about the use of force and they suggest a larger point. Some military organizations may jump on the opportunities for minimal force that the precision weaponry of the RMA allows in COIN. However, others, such as the US Army, are more likely to absorb these new means into its customary way of warfare that assumes the use of overwhelming firepower and the targeting of the adversary's military assets.

The importance of population security leads to an additional point about how the RMA may accentuate blind spots in the US Army's approach to COIN. The potential of the RMA to transform military organizations from large, heavy, and cumbersome forces to light, agile, and small networks could be advantageous to COIN. Light infantry formations are necessary for population security. However, the US Army's transformation is not, overall, creating a lighter force. The development of six Stryker brigades is the interim to the "Future Force" (or fully transformed army). These brigades are lighter and more rapidly deployable than the divisions that created such problems during the Kosovo operations. However, only one of the units chosen for conversion was a heavy brigade and it comes from the Reserves; the rest were previously light or medium units. In other words, the interim transformation of the US Army will lead to new medium units but the overall result will be a *heavier* army (Krepinevich 2004b: 49). The US Army's transformation will not result in a greater proportion of

COIN-capable forces; rather, it will retain a large percentage of units tailored to the use of decisive and overwhelming firepower, that is in conventional operations.

The RMA's lethality and the emphasis on standoff capabilities accord well with another article of faith in the US Army's culture: *casualty aversion* (Cassidy 2004b: 2; Caniglia 2001). This distinctive US approach became widely known during the Balkans operations and earned US forces the designation of "Ninja turtles." US troops patrolled in full combat gear with weapons, helmets, and body armor, and retired to isolated camps, cut off from the surrounding population. In contrast, European troops patrolled with weapons, but in small groups and wearing berets. They also were more likely to live, eat, and spend their free time in town. The Europeans do not care less about the lives of their soldiers than do Americans. This contrast reflects a deeper philosophical difference about the optimum approach to unconventional operations. While Americans assume that strength deters attacks and encourages cooperation, Europeans believe that a more relaxed and less aggressive posture engages with the population and builds goodwill and so reduces the threat of attack and potential casualties (Frankel 2003).

Given the US Army's approach to force protection, the development of standoff capabilities is likely to limit the Army's effectiveness in COIN. RMA proponents envision a battlefield practically emptied of soldiers: small units of networked forces will engage adversaries at standoff ranges, with little risk to themselves. "Seeing first, knowing first, and acting first" may be optimal in a conventional conflict focused on maximizing the destruction of an adversary's forces. But distancing one's forces from an adversary that mixes easily in and out of the wider population may undermine a "hearts and minds" campaign. Government forces must intensively patrol populated areas if the public is to gain confidence that they will be protected from insurgent retaliation. That assurance requires "boots on the ground." An officer in the British army explained: "You can't win hearts and minds from the back of an armored vehicle" (Frankel 2004). This "boots on the ground" approach is labor intensive, and runs counter to the RMA's promise to trade precision firepower for manpower and provide for a smaller but more lethal force (Cohen 1996; Metz 1997). As a British defense analyst said, "[British] commanders in Iraq last year said they didn't see how some swanky new technology would help. What helped was people on the ground patrolling" (Mathieson 2004). Nor can you gain public support or provide population security from undue reliance on standoff capabilities. Yet, it is this direction that the RMA is likely to push the US Army, especially given its casualty aversion. The use of standoff capabilities is likely to further isolate the US Army from the civilian population and so limit its effectiveness. In addition, although standoff weapons can be used to interdict outside support for an insurgency, this use risks distracting Army efforts from the most important source of insurgent strength: the local population (Metz and Kievit 1994).

The RMA also promises rapid victories, which is again an attractive prospect to a military stressing force protection, but will result in a military ill-equipped for COIN. This goal is sensible in conventional conflicts, but the more that the RMA encourages the US Army to think about and structure itself for short campaigns, the more ill-equipped it will be to combat an insurgency. Strategists concur that there are no quick victories in guerilla warfare. It took Mao's forces two decades to win China and the British Army needed 12 years to defeat the Malayan insurgency. Promising a quick victory with fewer troops to a military organization that already emphasizes force protection will only further hinder its recognition of the preparation necessary to combat an insurgency.

As with precision weaponry, the RMA's "battlespace awareness" may be relevant to COIN. Unmanned aerial vehicles such as the Global Hawk and Predator provide valuable information in Iraq and Afghanistan, and improved communications help avoid friendly fire and facilitate logistical support. Of course, insurgent forces will try to evade high-tech's prying eyes and ears. Technical surveillance systems are not a problem per se. The real problem is that RMA will encourage the Army – already *enamored with high-tech fixes* – to imagine a technical solution, at the expense of the development of human intelligence, to a problem that demands face-to-face contact (*Financial Times* 2005). When the adversary intermingles with, and is indistinguishable from, the population, the primary issue is figuring out intentions, and that requires human intelligence, not downloads from satellite imagery. Despite US information superiority, senior military officers acknowledge that their assessments of the Iraqi insurgency are often little more than guesses (Graham 2005; Pincus 2005).

This outcome seems especially likely in the US Army because technical intelligence also corresponds with its desire to minimize casualties, manifesting itself in standoff capabilities and search and destroy missions. The forces must actively patrol the region and build links with the population if the latter is to provide the human intelligence necessary to distinguish friend from foe. The Stryker vehicle, which relies heavily on information technology for protection (using information to avoid danger zones and thus not requiring heavy armor), reflects this over-reliance on technological fixes for human intelligence problems. Avoiding danger is all fine and good on a conventional battlefield, but knowing intentions with technology is difficult and staying away is counterproductive in counterinsurgencies. RMA innovations may lead US soldiers to become too reliant on a digitized battlespace that lacks critical information. Unfortunately, that is not all it is about. Defeating an insurgency requires human intelligence. Yet while the Air Force and Army focused on developing technical intelligence to combat the growing Iraqi insurgency, the US military neglected the most promising solution: working in partnership with Iraqi forces. The US military does not have the assets to develop its own human intelligence, yet for the first year of the war it did not work with Iraqi forces or focus the latter's training on counterinsurgency (Cordesman 2004: 14, 23–6; Hendren 2005). Blinded by a failure to see the war for what it is – an unconventional one – and drawn to possible technological solutions that could minimize casualties, the Army persisted in ignoring the path that could have led to the collection of useful human intelligence.

There is one additional belief in the US Army's culture that is important to this discussion: the US Army's narrow focus on the apparent "military" dimensions of warfare. The Vietnam experience reinforced this bias and it is reflected in the army's embrace of Samuel Huntington's conception of a sharp *military–political divide* (Adams 1990; Cassidy 2003). The conventional wisdom within the US Army is that the soldier should be isolated from politics, but given a free hand in military decisions. According to one student of US Army culture, "the US Army has embraced Clausewitz as the quintessential oracle of war, but it has also tended to distance itself from Clausewitz's overarching theme – the linkage of military instrument to political purpose" (Cassidy 2004a: 107).

This tendency to separate the military from the political is especially problematic in COIN. An army historian commented on the war in Iraq: "US war planners ... conceived of the war far too narrowly" and tended to think of stability operations "as someone else's mission" (Ricks 2004). This civil–military divide continued after Saddam's fall. US military commanders were reluctant to stop the looting that then fueled the resentment against the US occupation, and "tolerated, if not encouraged," poor relations between civil and military authorities. The military did not "see" the need for cooperation with the Provisional Authority. US commanders "focused on the military dimension of battle and forgot the fundamental principle that all victory is ultimately political in character" (Cordesman 2004: 9).

The contrast between the British and American armies illuminates this narrow American conception of what is "military." A retired British officer explained that

> [T]here's a warrior-wimp syndrome in the US Army. The Americans treat civil affairs [relations with local civilians] as a specialization and have specialized civil affairs battalions to do the touchy-feely stuff. [American] warriors stay as warriors and perceive themselves as warriors.

In contrast, in the British Army, "every single soldier has to become an agent of the civil affairs program ... we teach our young officers and soldiers all of this touchy-feely stuff right from the beginning" (Frankel 2003).

Recognizing the politics in war would help US forces understand that cultural sensitivity can be as important to COIN as precision weaponry. For the US Army, knowledge of the local population's culture and traditions merits attention only if time allows, once soldiers have learned "military" skills. While observing a soccer match between Royal Marines and Iraqi civilians, a former British officer likened current American attitudes to former colonial British attitudes: "We thought everyone wanted to be an Englishman and live an English life ... Americans seem to believe that everyone is envious and wants to be part of America." She then held up photos of US soldiers at a checkpoint ordering Iraqi civilians to lie on the ground.

> For Arabs, to be spread-eagled with your face in the dust is a hugely shameful thing. What the Iraqis want more than anything is respect, and to feel

valued and feel treated as human beings. What [British] troops show is a willingness to be humble, to lose a football match 9 to 3. If the Americans had played, I'm afraid they would have wanted to win.

(Frankel 2003)

The Americans probably would not have understood that their conduct in a soccer match had implications for the larger conflict.

This inattention to the politics of war echoes Vietnam. In his classic book on Vietnam, Andrew Krepinevich (1986: 131) explains that "the [US] Army remained convinced that the essence of the conflict was military, not political. Politics would have to take a back seat while the army inflicted damage on the insurgents to force them to the peace table." Yet politics cannot be postponed unless it is a war of annihilation. In COIN, socio–economic and political problems must be addressed, the use of overwhelming power avoided, and respect for the local population required. Yet, nothing about the RMA, when coupled to the US Army culture, points in that direction.

Changing an organization's culture

The previous section argues that important values and assumptions in the US Army's culture are likely to result in a situation where the Army seems well positioned to exploit the potential of the RMA at the high end of military operations with the forecast with COIN being less promising. If this assessment is correct, what is to be done? The analysis first considers whether we can expect an organization's culture to change naturally in response to functional and structural demands, and then focuses on the possibility of engineering changes in a military's culture. It closes with three policy recommendations.

Natural adaptation to external incentives?

Perhaps there is little reason for concern about the mismatch between the RMA, COIN, and US Army culture. Cultural change may be slow, but an organization's culture may adapt to changes in its environment. The longer COIN campaigns the US Army fights, the more suited for them its culture may become. However, four reasons suggest otherwise.

The first reason to doubt that the US Army's culture will adapt to changes in its environment is that *structure does not determine culture* – if it did, then there would be no need to distinguish theoretically between the two concepts. Culture is a system of meanings; social structure captures behavioral patterns (Alvesson 2003: 5; Meek 1998: 453). The same structure can accommodate different cultures: a study of industries using different production technologies in three countries found greater cultural similarities within each country than among firms with similar production processes (Maurice *et al.* 1980). We see a similar outcome with the military. Despite similar technology and organizational structures, their cultures often differ, even for organizations as apparently similar as the British, French and American armies (Soeters 2000: 468–70; Soeters and Recht 1998).

Different cultures can fit with the same structure and work well together, but this is not necessarily the case. Structural change may not lead to a corresponding evolution in basic values and assumptions. Indeed, this disjuncture between an organization's structure and culture generated much of the interest in organizational culture. Many scholars argue that attempts at organizational change often flounder because managers mistakenly assume that cultural change will follow structural changes (Pascale *et al.* 1997: 127–39; Johnson 1990: 183–200).

This is best seen in the US Army's cultural preferences and its actual historical experiences. The US Army's preference for big wars could reflect its involvement in the World Wars and its preparation to meet a Soviet assault along the Central Front. Yet that account ignores the role that small wars and nation-building have played in US military history, ranging from the conflicts against the Native Americans and the reconstruction in the South after the Civil War, to the engagements in the Philippines, Nicaragua, Germany, Japan, Greece, Vietnam, El Salvador, Haiti, Grenada, and the more recent involvement in Somalia, the Balkans, and Afghanistan. The US Army may view these missions as distractions from conventional warfare, but that belief does not accurately reflect its history. If culture changes in sync with experiences, then a readiness to engage in unconventional conflicts and play a governance role in post-conflict environments would be part of army culture. But that is not the case. Experiences do not lead directly to beliefs because beliefs interpret those experiences.

Second, an argument about *internal organizational adaptation* implies that actors examine evidence and then revise their beliefs based on this new evidence. The introduction of new technology or evidence of repeated failures would thus lead organizational members to rethink their orientation. Yet there are numerous examples of military organizations, such as the British Army during World War I and the US military in Vietnam, persisting to hold beliefs despite ample evidence to the contrary. If rethinking occurred, those officers advocating a change were marginal to, or subsequently marginalized from, the organization (Kier 1997: 101; Krepinevich 1986: 80). When organizational members rethink their beliefs, they use their pre-existing beliefs to interpret the new evidence rather than simply revise them based on new evidence (Jervis 1976). What organizational members see and the lessons they derive from organizational experiences reflect their cultures.

Pre-existing beliefs are likely to be especially influential in organizations, such as the military, that have strong cultures. If plausible alternative explanations exist, organizational members will gravitate to those that fit best with their beliefs. For example, the US Army's persistent application of conventional tactics in Vietnam led to its defeat, yet Army commanders did not learn – even in retrospect – the strategic futility of search and destroy missions or the value of population security (Krepinevich 1986). Instead, one of the apparent lessons of Vietnam was that the war was not *sufficiently* conventional: civilians had forced the military to fight with "one hand tied behind its back" (Betts 1996).

Privileging theory over data explains why experience can be "learning-irrelevant" outcomes (Tetlock 1998). Observers see what they already hold to be true, which means no one "learns" anything. As the previous example of the war-game for US military intervention in Iran illustrates, the disastrous consequences of ignoring post-conflict stability operations in Iraq has not changed beliefs: the US military is still envisioning that a future regime change operation would entail a rapid victory and withdrawal, with minimal stability operations (Fallows 2004). The US military continues to see constabulary missions as peripheral to its duties despite the accumulation of evidence to the contrary. Internal adaptation is also especially unlikely in the military because of the ambiguity of evidence. In contrast to private firms which have clear indicators of performance such as financial reports, much of the military's time is passed in peacetime where it receives little feedback on its effectiveness. Even in war, the feedback is not straightforward. Battlefield's complexity makes it difficult to determine the reasons for success or failure, and the greater the ambiguity, the more likely prior beliefs will affect how organizational members interpret their situation.

The military's *monopoly position* (and so the *absence of selection processes)* is the third reason to be skeptical that a military's culture will adapt to new demands. Population ecologists downplay internal adaptation and instead focus on external selection. They shift the focus from a single organization to "populations" of organizations and argue that fundamental change typically involves the replacement of one type of organization with another. In other words, the primary mechanism of organizational change is not an individual organization's purposeful adaptation but a Darwinian process of selective birth and death. The organizations that "fit" the environment survive; those that do not, drop out (Hannan and Freeman 1977: 929–66; Aldrich 1979). This process may explain both the apparent adaptation of some private sector firms to their environment and the failure of military organizations to do similarly.

The environment cannot select for the most appropriate organizations if there are no competitors. If an army does not adapt its values and beliefs to a new form of warfare, there is no auxiliary army ready to assume its place in the wake of military defeat. The US Army's inability to adapt to COIN in Vietnam did not lead to its replacement by another army with values and attitudes tailored to this challenge. The Army may have suffered a decline in resources and prestige, but another organization with a culture better suited to that environment did not replace it. Inter-service rivalry provides some competitive pressures on each service and especially between the US Army and Marine Corps on COIN missions. But the incentives are slight relative to a market environment. Armies may lose wars, but unlike firms, they rarely go bankrupt; they are not dismantled and their facilities sold to the highest bidder. Even armies that go "bankrupt," such as the French during World War II, are often reconstituted with few changes. This means that if cultural change occurs, it must be within the organization and not because a better-suited one replaces it. This requirement for organizational learning (rather than replacement) makes change unlikely. When a successful firm replaces a bankrupt one the selection process occurs automatically; no one has to "see" the reasons one firm is more successful than another or change the firm in the face of determined resistance. Not so with the military. Because military organizations must both determine the reasons for defeat and engineer the necessary changes within the organization, this adaptation is unlikely.

The *politics of organizational change* is the final reason that cultural change is unlikely. Organizational culture makes coordination possible, but culture is not neutral: some win and some lose by the values the organization upholds. Cultural change may generate opposition from vested organizational interests. The US war-game Millennium Challenge 02 illustrates this problem. Designed to test modern war-fighting developments, the game "validated" RMA concepts such as rapid, decisive operations. Yet according to the retired Marine officer who played the enemy commander, the game was scripted to ensure those results (Kaplan 2003). The ability of low-tech options to elude US technology was not the politically correct lesson. Or in the case of Vietnam, some US military officers and advisors recognized that the war should be fought differently, but they faced an uphill battle against powerful members of the organization. Even if organizational members see alternatives, political barriers may block their realization.

Engineering change in an organizational culture

If a military's culture is unlikely to transform naturally in response to changes in its environment, what about the possibility of directed change? Could a determined effort by leaders of the US Army transform aspects of Army culture? Answering this question requires returning to Smircich's distinction between organizational culture as something an organization *has* (a variable) or something an organization *is* (a process of enactment). The view one adopts critically affects advice about how – or if – one can change it.

Management consultants tend to view organizational culture as something an organization *has*, and they are optimistic about engineering cultural change. They see corporate culture as an adaptive regulating mechanism that managers can manipulate to promote organizational effectiveness. They also typically assume a cultural consensus throughout an organization (Deal and Kennedy 1982; Ouchi 1981; Peters 1993: 34–8). In contrast, pessimists – typically academics – tend to view organizational culture as something that an organization *is*, and are skeptical that leaders can be architects of cultural change. They are also likely to stress that cultures are ambiguous and often composed of numerous subcultures (Willmott 1993: 515–51; Legge 1994: 397–433; Pettigrew 1990; Handy 1985).

While practitioners profit handsomely from promises to change corporate cultures, the pessimists may have a more reasoned assessment of the possibility of planned change. This judgment stems partially from empirical studies which show that only 10–19 percent of attempts at corporate cultural change succeed

(Ogbonna and Harris 1998). But wariness about the possibility of cultural change also reflects the intellectual foundations of the practitioner's account. The optimists' see organizational culture as homogenous, and this eases the task of creating a new one, but it is rarely an accurate representation of organizational life. The greater the ambiguity and the more subcultures within an organization, the more complicated the cultural manipulation becomes. Not only is it difficult to figure out what the culture *is* that must be changed (because there are multiple ones), but management actions that shift one set of actors' beliefs one way, may have a different effect on members of a different subculture. An additional reason to question the practitioner's optimism stems from the basic nature of culture. Culture is a social construction, but its members often see its values and beliefs as objective reality; orientations that outsiders might see as just one possibility assume the position of received wisdom within the organization. Challenges to these "facts" are likely to meet resistance and arouse anxiety.

This discussion is not meant to imply that organizational cultures do not change. Of course they do. Social interaction continually makes and remakes the predominant values and beliefs within an organization. However, the issue is whether conscious cultural engineering is possible. Even the practitioners who argue that culture can be changed acknowledge that the process is difficult. Nonetheless, there is a middle position between the optimists and pessimists. The "realists" find some merit in each view and propose several conditions under which they expect managers to be able to influence (but not control) *cultural change* (Krefting and Frost 1985: 156). The key question here is whether military organizations will be able to exploit these opportunities. After all, few organizations have such strong cultures or control mechanisms and this difference may have implications for the potential for changing a military's culture.

Many organizational theorists argue that founders shape an organization's culture by creating the initial culture and choosing members, ceremonies, and procedures that reinforce it (de Vries and Miller 1986; Schein 1983: 13–28; Schein 1984; Selznick 1957; Martin *et al.* 1985). The dark suits, crisp white shirts, and narrow black ties of IBM employees were emblematic, for example, of the IBM founder's attempt to inculcate a distinctive culture (Watson 1963). This process can sometimes occur in the military: Admiral Hyman Rickover imposed his vision on the US Navy's Reactors Branch through his strong leadership and personal selection of officers who shared his focus on engineering excellence (Sims 2000: 65–79). However, if creating an organization's culture from scratch is easier than changing an existing one, there are few opportunities for this process to change a military's culture. Private firms are started every day, but there are three services in most militaries, and once established, they endure for generations. New subdivisions, such as the US Navy's Reactors Branch, provide opportunities for altering beliefs, but these occasions are costly, rare, and draw on existing organizational members.

Many organizational theorists emphasize the critical role of leadership in changing an organization's culture (Gagliardi 1986: 119; Smith 2003: 249–66;

Martin and Siehl 1983: 52–64; Schein 1985; Sims 2000). Leaders must unequivocally back the cultural change, clarify its objectives, persevere in supporting it, and model their behavior to reflect it. But the factor most often stressed is the "outsider" perspective of new executives (Martin and Siehl 1983). In contrast to current leaders who are as much a part of an organization's culture as its members, newcomers have escaped that acculturation. They also have greater emotional detachment from the organization's "way things are done." Yet again, this solution is rarely available to the military. Potential members must first pass a selection process that the military designs, and then are trained and educated by the organization itself. Few opportunities exist for "outsiders" to rise to the top.

The strength of a military's culture may be a double-edged sword. The consistency of beliefs means that most members have difficulty imagining that things could or should be done differently (Goffman 1961). Few leaders will "see" a need for change and many members will tenaciously resist challenges to their established truths (Gagliardi 1986: 119; Ogbonna 1993: 44). Yet the military's powerful assimilating mechanisms mean that once it commits to changing its culture, change may be more feasible than in a private company which has limited means to control and shape its members' beliefs. After all, the military can go beyond pronouncing new goals and redesigning the internal layout of their buildings; it can intervene at the lowest level to enforce behavioral compliance and indoctrinate its members (Sims 2000; Smith 2003). In short, although it may be harder for leading officers to overcome the initial hurdle of recognizing an alternative approach, the military may be relatively well-equipped to impose cultural change.

The military has one additional advantage that many organizations lack. Like any public institution, the military is subject to Congressional and executive oversight and this link to outside actors may help it bypass its cultural blinders. Members of the military may not "see" the need for change or imagine alternatives, but other parts of the government might. The forecast here is mixed. There are good reasons to be skeptical about the effectiveness of civilian intervention, at least in the US context. President Kennedy had from the onset directed high-level attention at developing counterinsurgency capabilities and he used his leverage over promotions and budgets to drive home his view. Yet the Army persisted in clinging to its preferred way of warfare (Krepinevich 1986: 27–32). This outcome should not come as a surprise: organizations guard their autonomy. But drawing lessons from the American experience may be unwise. Militaries in other states with different domestic political structures may be less capable of protecting their organizational autonomy and so more susceptible to civilian intervention (Avant 1994).

Conclusion

Changing US Army culture so that RMA's integration will not degrade the Army's ability to wage a counterinsurgency will be difficult. We should be especially wary of assertions about changing basic assumptions within the Army's culture so that the RMA will not degrade its ability to wage counterinsurgencies. On rare occasions the deepest level of a culture may be challenged, and changed, but few leaders would attempt or could carry that off. The most likely candidates for cultural change are less deeply-rooted aspects of a culture, such as artifacts that reflect a culture's beliefs or articulated values and beliefs (Schein 1984: 3–16). The more that a belief is so embedded that it is not expressed the less likely it is that it can be changed. This leads to the first recommendation: focus on accessible beliefs within the organization. The emphasis on casualty aversion in the US Army might be one possibility. This belief is strongly held, but also frequently articulated: it is not so deep-rooted that no mention needs to be made of it. In addition, this belief's institutionalization within the formal structure of the US Army (and thus routinization of it) could be reversed (Caniglia 2001). Finally an important corollary of this belief – the argument that the American public is unwilling to accept casualties – is more myth than reality. This argument misconstrues the historical record, and the empirical evidence could be used to challenge the Army's belief in the primacy of force protection (Feaver and Gelpi 1999; Burk 1999: 53–78; Larson 1996).

Given the difficulties of changing a culture, the best advice may be to work with it. This is the second recommendation. Actors constantly create and reproduce culture, but they do not do so in a vacuum: new ideas are more likely to succeed if they resonate with current values and beliefs (Heracleaous 2001; Krefting and Frost 1985: 157; Smith 2003: 249–62). In other words, those attempting to change aspects of a culture should be sufficiently removed from it to imagine alternatives, but they should also be close enough to it to understand it. That will often require an intensive study of the organization's culture. Theorists debate the relative value of qualitative and quantitative methods of discerning a culture's content, but regardless of the method adopted, this step must be taken (Lim 1995: 16–22; Heracleaous 2001; Martin 2002: 205–37; Ogbonna and Harris 1998; Schein 1999). Attempts to alter aspects of a culture must build on knowledge of it. The introduction of values and beliefs that are unknowingly antagonistic with the current orientation are more likely to lead to "culture shock" than mutual understanding.

Finally, there may be circumstances when shifts in beliefs or working within the culture are not sufficient. If drastic cultural change is necessary, what is to be done? One could try to change the organization's culture, but that must only be undertaken with an awareness of the difficulties entailed and uncertain prospects. The better path might be to create a new organization or build on an existing, but alternate one. For example, if the US military is to effectively combat insurgencies, it is probably best not to attempt to transform the US Army's culture within the artillery, infantry, and armored units, but instead to look elsewhere for those capabilities. US Special Forces might be expanded, or these capabilities might be developed in what are traditionally seen as non-combat units, such as military police units. These recommendations have problems – they are costly, disruptive, and may challenge vested interests – but they may provide the greatest prospect for success. The US Army's ability to fight one of its most likely contingencies may depend on it.

References

Adams, Thomas K. (1990) *Military Doctrine and the Organizational Culture of the U.S. Army*, PhD Dissertation, Syracuse University.

Aldrich, Howard E. (1979) *Organizations and Environments*, Englewood Cliffs, NJ: Prentice-Hall.

Allison, Graham T. and Kelley, Paul X. (2004) *Nonlethal Weapons and* Capabilities, Washington, DC: Council of Foreign Relations.

Alvesson, Mats (2003) *Understanding Organizational Culture*, Beverly Hills, CA: Sage.

Avant, Deborah (1994) *Political Institutions and Military Change: Lessons from Peripheral Wars*, Ithaca, NY: Cornell University Press.

Avant, Deborah (1996–97) "Are the Reluctant Warriors out of Control?" *Security Studies*, 6(2): 51–90.

Barley, R., Meyer, G.W., and Gash, D.C. (1988) "Cultures of Culture: Academics, Practitioners and the Pragmatics of Normative Control," *Administrative Science Quarterly*, 33 (1): 24–60.

Betts, Richard (1996) "The Downside of the Cutting Edge," *The National Interest*, 45 (Fall 1996): 80–3.

Burk, James (1999) "Public Support for Peacekeeping in Lebanon and Somalia," *Political Science Quarterly*, 114(1): 53–78.

Burke, Jason (2003) "British Tread Softly to Win the Peace," *Observer*, 5 September.

Burns, Robert (2004) "Marine Commander: Fallujah Offensive has 'Broken the Back' of the Insurgency," *Associated Press*, 18 November.

Campbell, Kenneth J. (1998) "Once Burned, Twice Cautious: Explaining the Weinberger–Powell Doctrine," *Armed Forces and Society*, 24(3): 357–74.

Caniglia, Richard R. (2001) "73 US and British Approaches to Force Protection," *Military Review*, 81(4): 73–81.

Cassidy, Robert M. (2003) "Prophets or Praetorians? The Uptonian Paradox and the Powell Corollary," *Parameters*, 33(3): 130–71.

Cassidy, Robert M. (2004a) "Back to the Street without Joy: Counterinsurgency Lessons from Vietnam and other Small Wars," *Parameters*, 34(2): 73–83.

Cassidy, Robert M. (2004b) *Peacekeeping in the Abyss: British and American Peacekeeping Doctrine and Practice after the Cold War*, Westport, CT.: Praeger.

Catanzaro, Michael (2001) "The 'Revolution in Military Affairs' has an Enemy: Politics," *The American Enterprise*, 12(7): 24–7.

Cebrowski, Arthur K. and Garstka, John J. (1998) "Network Centric Warfare: Its Origin and Future," *Proceedings of the Naval Institute*, 124(1): 28–35.

Cohen, Eliot A. (1996) "A Revolution in Warfare," *Foreign Affairs*, 75(2): 37–54.

Cordesman, Anthony (2004) *Strengthening Iraqi Military and Security Forces*, Washington, DC: Center for Strategic and International Studies.

Deal, Terence E. and Kennedy, Allan A. (1982) *Corporate Cultures; the Rites and Rituals of Corporate Life*, Reading, MA: Addison-Wesley.

Demchak, Chris C. (1996) "Numbers of Networks: social Constructions of Technology and Organizational Dilemmas in IDF Modernization," *Armed Forces and Society*, 23(2): 179–208.

Department of Defense (2003) *The Transformation Planning Guidance*, Washington DC: Department of Defense.

de Vries, Manfred F.R. Kets and Miller, Danny 1986) "Personality, Culture and Organization," *Academy of Management Review II*, 11(2): 266–79.

Eden, Lynn (2003) *Whole World on Fire: Organizations, Knowledge, and Nuclear Weapons Devastation*, Ithaca, NY: Cornell University Press.

Evans, Michael (2004) "Army Chief Admits Friction with U.S. Commanders", *The Times*, 21 April.

Fallows, James (2004) "Will Iran Be Next?" *Atlantic Monthly*, December 2004, 294(5): 99–110.

Feaver, Peter D. and Gelpi, Christopher (1999) "A Look at Casualty Aversion: How Many Deaths Are Acceptable? A Surprising Answer," *Washington Post*, 7 November, B3.

Financial Times (2005) "All Agree Insurgents are Overwhelming Domestic, Sunni, and Nationalist," 29 January, p.7.

Frankel, Glenn (2003) "British Troops Bring Their Brand Of Civic-Minded Peacekeeping To Iraq," *Washington Post*, 4 April, p.30.

Gagliardi, Pasquale (1986) "The Creation and Change of Organization Cultures: A Conceptual Framework," *Organization Studies*, 7(2): 117–34.

Goffman, Erving(1961) *Asylums: Essays on the Social Situation of Mental Patients and other Inmates*, NY: Anchor Books.

Graham, Bradley (2005) "U.S. to Pull 15,000 Troops Out of Iraq," *Washington Post*, 4 February, p.A01.

Grant, R. P. (2000): "The RMA Europe Can Keep in Step," European Union Institute for Security Studies Occasional Paper 15.

Handy, C. (1985) *Understanding Organizations*, Harmondsworth: Penguin.

Hannan, Michael T. and Freeman, John (1977) "The Population Ecology of Organizations," *American Journal of Sociology*, 82(5): 929–66.

Hendren, John (2005) "U.S. to Overhaul Training of Iraqi Forces," *Los Angeles Times*, 20 January.

Heracleaous, Loizos (2001) "An Ethnographic Study of Culture in the Context of Organizational Change," *The Journal of Applied Behavioral Science*, 37(4): 426–46.

Jehl, Douglas and Lichtblau, Eric (2004) "Review at C.I.A. and Justice Brings No 9/11 Punishment," *New York Times*, 14 September.

Jervis, Robert (1976) *Perception and Misperception in International Politics*, Princeton: Princeton University Press.

Johnson, G. (1990) "Managing Strategic Change: The Role of Symbolic Action," *British Journal of Management*, 1(4): 183–200.

Joint Chiefs of Staff (1996) *Joint Vision 2010: America's Military Planning for Tomorrow*, Washington, DC: Office of the Joint Chief of Staff.

Kaplan, Fred (2003) "War-Gamed: Why the Army Shouldn't be so Surprised by Saddam's Moves," *Slate.msn.com*, 28 March.

Kier, Elizabeth (1997) *Imagining War: French and British Military Doctrine Between the Wars*, Princeton, NJ: Princeton University Press.

Krefting, Linda A. and Frost, Peter J. (1985) "Untangling Webs, Surfing Waves, and Wildcatting: multiple-metaphor perspective on managing culture," in Peter J. Frost, Larry F. Moore, Meryl Reis Louis, Craig C. Lundberg and Joanne Martin (eds) *Organization Culture*, Beverly Hills, CA: Sage.

Krepinevich, Andrew F. (1986) *The Army and Vietnam*, Baltimore: Johns Hopkins University Press.

Krepinevich, Andrew F. (1999/2000) "Why No Transformation," *Joint Force Quarterly*, Autumn/Winter 1999–2000, 23: 97–101.

Krepinevich, Andrew F. (2003) *The Thin Green* Line, Washington, DC: Center for Strategic and Budgetary Assessments.

Krepinevich, Andrew F. (2004a) *The War in Iraq: The Nature of Insurgency Warfare*, Washington, DC: Center for Strategic and Budgetary Assessment.

Krepinevich, Andrew F. (2004b) *Transforming the Legions: The Army and the Future of Land Warfare*, Washington, DC: Center for Strategy and Budgetary Assessment.

Larson, Eric V. (1996) *Casualties and Consensus: The Historical Role of Casualties in Domestic Support for U.S. Military Operations*, Santa Monica: Rand.

Legge, Karen (1994) "Managing Culture: Fact or Fiction," in Keith Sisson (ed.) *Personnel Management: A Comprehensive Guide to Theory and Practice in Britain*, Oxford: Blackwell.

Legro, Jeffrey W. (1995) *Cooperation under Fire: Anglo-German Restraint during World War II*, Ithaca, NY: Cornell University Press.

Lim, Bernard (1995) "Examining the Organizational Culture and Organizational Performance Link," *Leadership and Organization Development Journal*, 16(5): 16–22.

Lyons, James (2004) "Attack on Fallujah Won't Work, Says Cook," *Press Association*, 10 November.

Martin, Joanne (1992) *Cultures in Organizations: Three Perspectives*, New York: Oxford University Press.

Martin, Joanne (2002) *Organizational Culture: Mapping the Terrain*, Beverly Hills, CA: Sage.

Martin, Joanne and Siehl, Caren (1983) "Organizational Culture and Counter-Culture: An Uneasy Symbiosis," *Organizational Dynamics*, 12(2): 52–64.

Martin, Joanne, Sitkin, Sim B., and Boehm, Michael (1985) "Founders and the Elusiveness of a Cultural Legacy," in Peter J. Frost, Larry F. Moore, Meryl Reis Louis, Craig C. Lundberg and Joanne Martin (eds) *Organizational Culture*, Beverly Hills CA: Sage.

Mathieson, S.A. (2004) "At the Ready: The Ministry of Defence is using Technology to Reduce Troops – but Could it Endanger Peacekeeping Operations?" *Guardian*, 9 September.

Maurice, Marc, Sorge, Arndt and Warner, Malcolm (1980) "Societal Differences in Organizing Manufacturing Units: A Comparison of France, West Germany, and Great Britain," *Organisational Studies of Management and Organisation*, Vol. 10, No. 4.

Meek, Lynn (1998) "Organizational Culture: Origins and Weaknesses," *Organization Studies*, 9(4): 453–73.

Metz, Steven (1997) "Racing Toward the Future: The Revolution in Military Affairs," *Current History*, 96: 184–8.

Metz, Steven and Kievit, James (1994) *The Revolution in Military Affairs and Conflict Short of War*, Carlisle, PA: US Army War College, Strategic Studies Institute.

Mockaitis, Thomas (1990) *British Counter-Insurgency in the Post Imperial Era*, Manchester University Press.

Morris, Nigel (2004) "Iraq in Chaos: British General Admits Anglo-U.S. Friction," *Independent*, 21 April, p.24.

Ogbonna, Emmanuel (1993) "Managing Organizational Culture: Fantasy or Reality," *Human Resource Management Journal*, 3(2): 42–54.

Ogbonna, Emmanuel and Harris, Lloyd C. (1998) "Managing Organizational Culture: Compliance or Genuine Change?" *British Journal of Management*, 9(4): 273–88.

Ouchi, W. (1981) *Theory Z*, Reading: Addison-Wesley.

Pascale, R., Milleman, M., and Gioja, L. (1997) "Changing the Way We Change," *Harvard Business Review*, Nov./Dec. 1997 75(6): 127–39.

Peters, Thomas J. (1993) "On Culture," *Management Decision*, 31(5): 34–8.

Peters, Thomas J. and Waterman, Robert H. (1982) *In Search of Excellence: Lessons from America's Best-run Companies*, New York: Harper & Row.

Pettigrew, A. (1990) "Is Corporate Culture Manageable?" in D. Wilson and R. Rosenfield (eds) *Managing Organizations*, London: McGraw Hill.

Pincus,Walter (2005) "CIA Studies Provide Glimpse of Insurgents in Iraq," *Washington Post*, 6 February, p.A19.

Ricks, Thomas E. (2004) "U.S. Military Not Plan for Post Conflict, Stability Operations in Iraq: Army Historian Cites Lack of Postwar Plan," *Washington Post*, 25 December.

Rudebeck, Clare (2003) "Charm Offensive: the Army's Secret Weapons for Iraqi Hearts," *Independent*, 3 October, p.18.

Schein, Edgar (1983) "The Role of the Founder in Creating Organizational Culture," *Organizational Dynamics*, 12(1): 13–28.

Schein, Edgar (1984) "Coming to a New Awareness of Organizational Culture," *Sloan Management Review*, 25(2): 3–16.

Schein, Edgar H. (1985) *Organisational Culture and Leadership: A Dynamic View*, San Franciso, CA: Jossey-Bass.

Schein, Edgar (1992) *Organizational Culture and Leadership*, San Francisco: Jossey-Bass.

Schultz, M. and Hatch, M.M. (1996) "Living with Multiple Paradigms: The Case of Paradigm Interplay in Organizational Culture Studies," *Academy of Management Review*, 21: 529–57.

Selznick, Philip (1957) *Leadership in Administration: A Sociological Interpretation*, New York: Row, Peterson.

Short, Anthony and Coates, John (1992) *Suppressing Insurgency: An Analysis of the Malayan Emergency, 1948–1954*, Boulder, CO: Westview Press.

Sims, Ronald R. (2000) "Changing an Organization's Culture under New Leadership," *Journal of Business Ethics*, 25(1): 65–79.

Singer, Peter W. (2003) *Corporate Warriors: The Rise of the Privatized Military Industry*, Ithaca, NY: Cornell University Press.

Smircich, Linda (1983) "Concepts of Culture and Organizational Analysis," *Administrative Science Quarterly*, 28: 339–58.

Smith, Martin E. (2003) "Changing an Organization's Culture: Correlates of Success and Failure," *Leadership and Organizational Development Journal*, 24(5/6): 249–66.

Smith, Simon C. (2001) "General Templer and Counter-Insurgency in Malaya: Hearts and Minds, Intelligence, and Propaganda," *Intelligence and National Security*, 16(3): 60–78.

Soeters, Joseph L. (2000) "Cultures in Uniformed Organizations," in Neal M. Ashkanasy, Celeste P.M. Wilderom, and Mark F. Peterson (eds) *Handbook of Organizational Culture and Climate*, Thousand Oaks: Sage Publications, 2000.

Soeters, Joseph L. and Recht, R. (1998) "Culture and Discipline in Military Academies: An International Comparison," *Journal of Political and Military Sociology*, 26(2): 169–89.

Stubbs, Richard (1989) *Hearts and Minds in Guerrilla Warfare: The Malayan Emergency, 1948–1960*, Singapore: Oxford University Press.

Tetlock, Phillip (1998) "Social Psychology and World Politics," in Daniel T. Gilbert,

Susan T. Fiske and Gardner Lindzey (eds) *Handbook of Social Psychology*, NY: McGraw Hill.

The Economist (2003) "Hearts and Minds in Southern Iraq," 5 April.

The Economist (2005) "When Deadly Force Bumps into Hearts and Minds," 1 January, p.40.

Thornton, Rod (2004) "The British Army and the Origins of its Minimum Force Philosophy," *Small Wars and Insurgencies*, 15(1): 83–106.

Watson, Thomas J. Jr. (1963) *A Business and its Beliefs: The Ideas that Helped Build IBM*, New York: McGraw-Hill.

Wilderom, Celeste P.M., Glunk, Ursula and Maslowski, Ralf (2000) "Organizational Culture as a Predictor of Organizational Performance," in Neal M. Ashkanasy, Celeste P.M. Wilderom, and Mark F. Peterson (eds) *Handbook of Organizational Culture and Climate*, Thousand Oaks: Sage Publications.

Williams, Michael (1998) *Civil Military Relations and Peacekeeping*, London: International Institute for Strategic Studies.

Willmott, H. (1993) "Strength is Ignorance; Slavery is Freedom: Managing Culture in Modern Organizations," *Journal of Management Studies*, 30(4): 515–51.

10 Officer attitudes toward the revolution in military affairs

Thomas G. Mahnken and James R. FitzSimonds

For at least a decade, a chorus of defense analysts, government officials, and military officers have argued that the growth and diffusion of stealth, precision, and information technology will drastically alter the character and conduct of future wars, yielding a revolution in military affairs (RMA). The idea that the emergence of new technology, combined with innovative operational concepts and organizations, would transform the conduct of war, first appeared in Soviet military writings in the late 1970s (Goure and Deane 1984). It was, however, the seeming ease with which the United States-led coalition defeated Iraq during the 1991 Gulf War that led many observers in the United States (US) and elsewhere to conclude that significant changes in the character of warfare were underway (Cohen 1996). Since the mid–1990s, exploiting the emerging RMA has been an explicit goal of the Department of Defense (DoD). Each of the Services has devoted considerable attention to developing new technology, as well as the concepts and organizations needed to employ it effectively.

This chapter presents selected results of the first systematic effort to understand officer attitudes toward military transformation. It is based upon surveys conducted in 2000 and 2002 of more than 4,500 officers attending US professional military education (PME) institutions. It shows that the military transformation agenda has generated mixed emotions amongst the US military's officer corps. On the one hand, there is significant support for the need to transform the US armed forces based both on emerging threats and opportunities. There is also strong support for organizational change, at least in the abstract. On the other hand, the research reveals that many officers perceive barriers to innovation within their services. Moreover, many officers are uncertain whether the US armed forces are currently on the right path to transformation.

Transforming the US armed forces

George W. Bush campaigned on a pledge to transform the US armed forces by skipping a generation of technology. In a September 1999 campaign speech at the Citadel military college, then-governor Bush noted, "our military is still organized more for cold war threats than the challenges of the new century – for industrial – age operations, rather than information-age battles" (*New York*

Times 24 September 1999). Transforming the US armed forces became one of the Bush administration's top priorities when it took office. Speaking at the Norfolk Navy Base in February 2001, President Bush promised to "move beyond marginal improvements to harness new technologies that will support a new strategy" (*New York Times* 14 February 2001). He called for the development of ground forces that are lighter, more mobile, and more lethal, as well as manned and unmanned air forces capable of striking across the globe with precision.

Soon after assuming office, Secretary of Defense Donald Rumsfeld commissioned Andrew W. Marshall, the Pentagon's premier strategic thinker, to conduct a fundamental review of US strategy and force requirements. He also commissioned a panel of senior experts to develop a transformation strategy for the Pentagon (*Aerospace Daily*, 29 August 2001). However, early proposals to reduce the size of the US armed forces and cancel major acquisition programs to fund the development of new weapon systems generated opposition among members of Congress and senior members of the armed services (*Washington Post* 20 May 2001). The DoD's 2001 Quadrennial Defense Review (QDR) contained none of the radical changes that had originally been discussed within the DoD.

However, the war on terrorism has given transformation a new lease of life. In a second speech at the Citadel on 11 December 2001 (Office of the Press Secretary 2001), President Bush renewed his call for the transformation of the US armed forces. Arguing that "the conflict in Afghanistan has taught us more about the future of our military than a decade of blue ribbon panels and think-tank symposiums," Bush called upon the military to field forces that would rely more heavily on unmanned air vehicles and precision-guided munitions. He also called for sacrifice, warning that

> [E]very service and every constituency of our military must be willing to sacrifice some of their pet projects. Our war on terror cannot be used to justify obsolete bases, obsolete programs, or obsolete weapons. Every dollar of defense spending must meet a single test: It must help us build the decisive power we will need to win the wars of the future.

In April 2003 the Bush administration issued its *Transformation Planning Guidance*, a document that was intended to outline the DoD's strategy for transforming the military services. The rapid US military victory in Iraq in early 2003 notwithstanding, the *Transformation Planning Guidance* (Department of Defense 2003) cites five reasons why major changes to US forces are now needed. One, the diffusion of information-age technology will erode US military advantages. Two, asymmetric threats are growing, including terrorism and weapons of mass destruction. Three, new force-on-force challenges are emerging, such as new electronic and cyber warfare capabilities, new means to counter or negate US advantages such as space capabilities, and anti-access capabilities such as submarines, mines and cruise and ballistic missiles. Four, there is a historic opportunity resulting from the US victory in the Cold War. Finally, transformation involves high stakes.

Why study officer attitudes?

There are four compelling reasons why it is important to understand officer attitudes toward military transformation. First, the military services will be the ultimate practitioners of new ways of war. The extent to which their members are enthusiastic about change may help determine the success or failure of new technologies, operational concepts, and organizations. Second, although very few officers will likely emerge as true innovators, the existence of a climate conducive to innovation within the officer corps may encourage individuals both to generate new ideas, and to remain in the service to bring them to fruition. Third, a large percentage of career officers will rise to senior leadership positions within their services in the next ten to 20 years. In those roles, they will establish command climates that will either support or inhibit risk-taking and innovation. Past research has demonstrated the importance of innovation to senior officers who protect and nurture the careers of young innovators under their command who are willing to take risks (Rosen 1991: 20, 251). Finally, officers are the recognized experts in military affairs in the US. They should be expected to take a leading role in determining the need for adopting different approaches to warfare.

Although officer attitudes may play an important role in the process of innovation, to date they have received little scrutiny. This chapter presents the results of a multi-year project designed to redress this shortfall (Mahnken and FitzSimonds 2003a; Mahnken and FitzSimonds 2003b). Since 2000, the project has employed two large-scale surveys and a series of focus groups to gain a better understanding of officer attitudes and what shapes them. First, between March and October 2000 the project included a survey of more than 1,900 officers attending seven US PME institutions. These are: the Naval War College (including the College of Naval Command and Staff, and College of Naval Warfare), Air Command and Staff College, Air War College, Army Command and Staff College, Army War College, National War College, and National Defense University's Capstone Course for newly-promoted flag officers. The survey population included junior and field-grade officers (Army, Air Force, and Marine Captains and Majors, Navy Lieutenants and Lieutenant Commanders), senior officers (Army, Air Force, and Marine Lieutenant Colonels, Colonels and Navy Commanders and Captains), and flag officers (Generals and Admirals) from all branches of the US military, their reserve components and National Guard, as well as foreign officers and US government civilians (Mahnken and FitzSimonds 2003b: 11–13). The project also included a series of focus groups to help to generate a better understanding of the results of the survey.

Second, in September and October 2002 a web-based survey of more than 2,500 officers – ranging from O–3s to O–7s – attending 14 PME institutions – including the services' war colleges and command and general staff colleges, the

Naval Postgraduate School as well as the National War College. The survey population included 2,147 US and 188 foreign officers. It included 962 Army, 296 Navy, 877 Air Force, and 191 Marine officers. This survey included many statements that had appeared in the 2000 survey as well as a number of new ones.[1] Although the survey population may not be a characteristic cross-section of the entire officer corps, it is representative of the subset of the officer corps that gets an opportunity to attend PME institutions. Army, Air Force, and Marine Corps officers in particular are selected to attend PME institutions based upon their potential for higher command. Responses from today's senior and flag officers offer insights into the attitudes of those who will be responsible for making decisions about how the armed forces transform themselves over the next five to ten years. By contrast, today's junior officers will occupy the leadership of the US armed forces in 2020–2025.

The study demonstrates that nearly two decades after the Goldwater-Nichols Act, which was designed to make the US armed forces more "joint," service affiliation remains the strongest determinant of officer attitudes. In general, officers' service affiliations proved more important in shaping attitudes than their war fighting specialties. In other words, the attitudes of infantry officers in the Army were more like those of Army officers as a whole than infantry officers in the Marine Corps. Rank also influenced attitudes, although to a much lesser degree. Senior officers were generally more expectant of change than junior and field grade officers. Other demographic variables, like whether an officer had served in combat, had little to no effect upon the responses.

Why transform?

US DoD statements contain two broad rationales for transformation. The first is that transformation is required to maintain competitiveness in the face of increasingly sophisticated adversaries. The second is that the information technology revolution offers opportunities to fight wars more effectively. The results of the surveys in this study demonstrated significant and growing support for large-scale change within the US armed forces. In 2000, for example, 47 percent of the officers surveyed agreed with the proposition that the US armed forces must radically change their approach to warfare to compete effectively with future adversaries. In 2002, by contrast, 57 percent agreed with the need for radical change. Perhaps just as significant, the percentage of officers who disagreed with the need for radical change dropped from 41 percent to 28 percent. However, nearly two-thirds tended toward uncertainty, suggesting that they were either unsure as to what was being done, or what exactly "radical" change entailed.

Not all services viewed the need for change with the same urgency. A majority of Army, Navy and Air Force officers – broken down to 58 percent of Army and Navy officers and 56 percent of Air Force officers, but only 42 percent of Marine Corps officers – agreed with the need for radical change; by contrast, only a minority of Marine Corps officers saw such a need. The breakdown among junior and senior officers is also revealing. More senior offices (59 percent) agreed that the US armed forces must change "radically" than junior officers (55 percent). While we can only speculate as to the disparity of these views, they may be the result of the fact that senior officers are more aware of the challenges facing the US armed forces than their juniors.

Respondents to the 2002 survey showed far greater concern over specific threats than did their counterparts in 2000. Moreover, those officers who were most concerned strongly supported changes to the US force posture to redress these threats. In short, they accepted one of the core assumptions behind transformation. It is not clear whether this was driven by real world events such as the 11 September 2001 terrorist attacks, or by Bush administration efforts to promote the need for change.

Most officers appeared to agree with the need to deploy forces that are less dependent upon ports and airfields. Of the officers surveyed, 69 percent believed that within the next ten years some adversaries would likely have the ability to use long-range precision strike weapons such as ballistic and cruise missiles to deny the US the use of fixed military infrastructure, such as ports, airfields, and logistical sites. This represented a complete turnaround from 2000, when only 9 percent agreed with the statement.[2] There was a consensus on this point; at least two-thirds of officers from each service – broken down into 73 percent of Army, 69 percent of Navy, 67 percent of Air Force, and 75 percent of Marine Corps officers –agreed with the statement. Moreover, 74 percent of senior officers agreed, compared to 68 percent of junior officers. While it is impossible to ascertain the source of this change, it appears that the 11 September 2001 terrorist attacks on the World Trade Center and Pentagon shattered a sense of invulnerability that had previously grown up.

Moreover, 82 percent of those who agreed felt the risk to forward bases of being attacked by precision strike weapons would force the US armed forces to introduce new operational concepts that would allow them to project power without reliance on forward bases.

Many officers also felt that the US armed forces would need to disperse in future conflicts. 46 percent believed that within the next ten years, the proliferation of long-range precision strike weapons would make it too risky for the US military to mass forces geographically in small areas as it did during the 1990–1991 Gulf War. Of those who agreed, 76 percent believed that this increased risk would require the US military to adopt lighter and more mobile forces, and new concepts of operations, in order to avoid being attacked.

Supporters of military transformation also argue that the information technology revolution is opening up new possibilities for the US military. According to the *Transformation Planning Guidance* (2003: 4), US "strategy requires transformed forces that can take action from forward position, and, rapidly reinforced from other areas, defeat adversaries *swiftly* and *decisively* while conducting an active defense of US territory."

Most officers appear to share this view. A large and growing majority believed that information-age ways of war would make it easier for the United

States to use force to achieve decisive battlefield victories with substantially reduced risk of US casualties. 70 percent of the officers felt that new technology, operational concepts, and organizations would make it easier to use force, 7 percent more than in 2000. Moreover, while 24 percent disagreed with the statement in 2000, only 18 percent disagreed in 2002.

An even larger majority of officers – 79 percent – felt that new ways of war would make it easier to achieve decisive battlefield victories, compared to 60 percent in 2000. Although it is difficult to tell for certain what caused this shift, it appears likely that the relative ease with which US forces dislodged the Taliban regime in Afghanistan influenced officers' attitudes toward the future. It is possible that the ongoing violence in Afghanistan and Iraq may have subsequently tempered this view.

Most officers surveyed in 2000 and 2002 believed that new ways of war are making combat less lethal for American servicemen. 69 percent felt that new technology, operational concepts, and organizations would offer the US the ability to engage in high-intensity operations with substantially reduced risk of casualties, 6 percent more than in 2000.

The organizational dimension

Survey respondents also reflected strong support for organizational change in the abstract. 60 percent of officers in 2000 and 59 percent of officers in 2002 believed that modern conditions require significant changes to traditional service roles and missions. Perhaps more significantly, resistance to such changes decreased between 2000 and 2002; while 32 percent of officers believed in 2000 that significant changes to roles and missions were not warranted, in 2002 that had fallen to 25 percent. Here, the dichotomy between Army, Navy, and Air Force attitudes, on the one hand, and Marine Corps attitudes, on the other, was apparent; while a majority of Army, Navy and Air Force officers – 65 percent of Army, 55 percent of Navy and Air Force – agreed with the need for change, little more than a third of Marines (35 percent) agreed. It was unclear whether officers supported change to service roles and missions in the abstract, or whether they saw such changes as benefiting (or hurting) their service.

Support for changes in service roles and missions did not carry over to support for a reduction in service autonomy. Moreover, the belief in the need for separate services has remained steady over time. 37 percent of officers in 2000 and 36 percent of officers in 2002 believed that the need to maintain separate services would diminish over time. By contrast, 52 percent of officers in 2000 and 51 percent of officers in 2002 disagreed. In other words, while service roles and missions may change, there is little support for service unification.

Of note, senior officers were more receptive than juniors to changes in the status quo, believing more strongly in the need for significant changes to service roles and missions and that the need for separate services would diminish over time. Perhaps senior officers are more open to change because they have seen significant changes during their careers.

Most officers were clearly open to the idea of major changes in dominant military skills. For example, in 2002, 57 percent believed that an individual in his branch or designator would likely need very different skills in 2020 compared to those required at the time of the survey. However, most officers did not believe that their own specialty could be rendered obsolete: only 18 percent believed that their service specialty could be rendered obsolete due to technological developments over the next 20 years.

These responses may reflect an officer's social identification with a specific branch or designator that is largely independent of a specific skill. For example, officers have identified themselves as members of the "cavalry" regardless of whether the branch was formed around horses, armored vehicles, or helicopters. This suggests that opposition to organizational change might be stronger if it seeks to move officers into new branches rather than seeking to change the dominant skill sets within an existing branch.

Nevertheless, there were a number of primary service specialties in which a relatively large percentage of respondents did feel that significant change was likely. The strong response from Air Force navigators (49 percent of whom felt that their specialty could be rendered obsolete) and Naval Flight Officers (47 percent) undoubtedly reflects trends toward drawdowns in those specific specialties.

The survey uncovered significant support for jointness. Nearly two-thirds of the officers surveyed considered increasing joint command authority at the expense of individual service autonomy to be a positive development. It is interesting that senior officers showed a slightly higher receptivity than junior officers toward increasing jointness – perhaps reflecting a more positive attitude stemming from more joint experience. Support for jointness was hardly uniform, however. Whereas a large majority of Army, Navy, and Air Force officers (69 percent, 63 percent, and 69 percent, respectively), considered increasing joint command authority to be a positive development, only 37 percent of Marines shared that view.

The cultural dimension

The Defense Department's *Transformation Planning Guidance* sets out a broad array of goals, including the transformation of what might be termed the military culture. As the document puts it,

> We must transform not only our armed forces, but also the Department that serves them by encouraging a culture of creativity and prudent risk-taking. We must promote an entrepreneurial approach to developing military capabilities, one that encourages people to be proactive, not reactive.
>
> (DoD 2003: 1)

The survey conducted in 2002 appears to indicate that most officers have little first-hand experience with innovation. Of the officers surveyed, 66 percent

Table 10.1 "Fear of failure inhibits true innovation in my branch of service"

Service	Branch	Disagree	Agree
Army	Armor	29	56
	Aviation	17	63
	Engineers	32	51
	Infantry	36	49
	Special forces	49	23
Air Force	Navigator	42	40
	Pilot	38	43
	Space	17	64
Navy	NFO	39	45
	Pilot	27	61
	Submarines	31	66
	Surface	37	48
Marines	Artillery	44	48
	Aviation	47	35
	Infantry	58	26

were uncertain whether their service tends to reward innovators, indicating either that they had not experienced innovation or did not know what happened to those innovators they encountered. There was a marked distinction between the services in response to the statement – with a clear majority of Air Force and Marine Corps respondents (56 percent and 58 percent, respectively) believing that their services did indeed reward innovators. By contrast, only 34 percent of Army and 28 percent of Navy officers agreed with the statement.

Nearly half (48 percent) of the officers surveyed believed that fear of failure inhibited innovation in their service branch. Junior officers held a decidedly more negative opinion than senior officers on this issue; half felt fear of failure inhibited true innovation in their branch, compared to 43 percent of senior officers. There was also a major variance among the services, with the Marines holding a significantly more positive view than their counterparts in the other services. Fully half of Marine officers disagreed with the contention that fear of failure inhibits true innovation in their branch, compared to 32 percent of Army, 34 percent of Navy, and 36 percent of Air Force officers. There were significant distinctions between the various specialties within each of the services (see Table 10.1).

On the positive side, a large percentage of officers (48 percent) expressed confidence that their own service branch had a culture that was open to self-criticism. Senior officers tended to have a slightly more positive view of their own service's propensity for critical self-analysis than did junior officers: 51 percent of senior officers believed their service branch had a culture that was open to self-criticism, compared to 48 percent of junior officers. There were also significant differences in responses among the services, with the Marines holding a much more positive view of service propensity for self-criticism compared with the Army, Navy, and Air Force. While 75 percent of the Marines we surveyed believed that their service had a culture that is open to self-criticism, a minority of 48 percent of Army and Air Force officers and 39 percent of Navy officers agreed.

Are the US armed forces on the right path?

As noted above, a majority of officers felt that the US armed forces must radically change to compete effectively with future adversaries. However, they were much less certain whether the US armed forces were doing so. Fully two-thirds of the officers surveyed were uncertain that the US armed forces were embarked upon a path that would lead to radical change in military technology, doctrine, and organization; 49 percent felt they were. Significantly, this was little changed from 2000, when 71 percent were unsure and 48 percent felt they were. This would seem to indicate that, in this area at least, the Bush administration's emphasis on transformation has yet to affect officer attitudes.

The study found significant differences among the services as to whether the US armed forces were on a path to radical change. 61 percent of Army officers, but only 36 percent of Navy, 48 percent of Air Force, and 28 percent of Marine officers believed that the armed forces were on the path to radical change. The strong Army response is likely due to the emphasis the service has placed on transformation in recent years. By contrast, Marines were most skeptical: 52 percent of Marine officers, but only 24 percent of Army, 43 percent of Navy, and 29 percent of Air Force officers disagreed with the statement.

Respondents expressed a generally negative view – or a least a very uncertain view – of what they saw happening compared with their views of what "real" transformation required. 80 percent felt that real transformation would require major changes to personnel management policies and procedures, but only 32 percent saw evidence of major changes to personnel management in their service. Moreover, 58 percent tended toward uncertainty. Similarly, 82 percent felt that real transformation would require major changes to military training and education, but only 41 percent saw evidence of major changes while 65 percent were unsure. There was no major difference between junior and senior respondents on these issues.

More than eight in ten officers believed that it was important for them to spend time thinking about change. Yet barely more than one-third believed that they had time to think seriously about the impact of advanced technology. There was virtually no distinction between junior and senior officers on these points. Service responses were also quite consistent.

The study uncovered a widespread expectation that the information technology revolution would lead to more centralization. Of officers surveyed, 65 percent expected to see more centralized control over US military operations in coming years; only 20 percent did not. However, these officers had an overwhelmingly negative view of centralization: nearly six in ten viewed it to be a negative development. Junior officers had a higher expectation, but a more negative view, of the trend toward centralization than senior officers. 66 percent

of junior officers and 62 percent of senior officers expected to see more centralized control over US military operations in the coming years; whereas 61
percent of junior and 59 percent of senior officers considered more centralization to be a negative development. While 84 percent of Marine officers viewed
centralization negatively, one-third of Air Force officers viewed it positively.
This appears to be a reflection of Air Force doctrine, which emphasizes the centralized control of air assets through a Joint Forces Air Component Commander.

These statements offered one of the more intriguing set of responses –
begging the question of what officers perceived to be the driver behind what
they see as an increasing trend toward centralization. On the one hand, they may
have perceived increasing centralization as an inexorable feature of the increasing exploitation of information technology, and one that no leader would be able
to stem. On the other hand, they might have seen centralization as the result of
conscious decisions being made by senior military leaders, thus calling into
question the confidence that officers have in what their senior leadership was
doing.

Observations

This study contains some good news for supporters of transformation within the
US DoD. It documents significant latent support for transformation, driven both
by perceptions of emerging threats and opportunities. A majority of officers
appear to accept the need to transform the US armed forces to reduce their vul-
nerability. A significant majority also believed that new ways of war would
allow the US to achieve decisive battlefield victories with substantially reduced
risk of US casualties. It also shows some support for organizational change,
albeit within the context of current service specialties.

Not all the news is so good, however. The study indicates that most officers
do not appear to have a lot of experience with innovation. Indeed, they felt that
some aspects of their service and branch's culture inhibited innovation. Fear of
failure was one common complaint.

What explains this pattern of responses? An officer's service affiliation was
the single most important determinant of his attitudes. Nearly 20 years after the
Goldwater–Nichols Act, which was designed to make the US armed forces more
"joint," service affiliation remains the strongest determinant of officer attitudes
that we could identify.

An officer's branch or specialty also affected his attitudes. However, an
officer's service affiliation appears to be a more important determinant of his
attitudes than his specialty or rank. A comparison of Army and Marine infantry
officers is illustrative. 58 percent of Army officers, and 57 percent of Army
infantry officers, believed that "the US armed forces must radically change their
approach to warfare to compete effectively with future adversaries." By contrast,
42 percent of Marine officers but only 30 percent of Marine infantry officers
agreed with the statement. Similarly, 65 percent of Army officers, and 65
percent of Army infantry officers, believed that "modern conditions require

significant changes to traditional Service roles and missions." By contrast, only 35 percent of all Marine and 14 percent of Marine infantry officers agreed. In other words, Army and Marine infantry officers accentuated the differences between the two services.

Senior officers appeared to be more open to change than junior officers. Out of 82 survey items related to transformation, the study found that junior officers were supportive in 17 cases, but senior officers were supportive in 32 cases. It appears that because senior officers have served longer in the military than their juniors, they have had an opportunity to witness more change throughout their careers. They also have much greater experience outside their branch or service than junior officers. Junior officers, by contrast, have a narrow base of experience upon which they can draw. While they are experts in their specialty, they have little experience outside their branch or community. Moreover, most are quite naturally concerned with promotion and force command, opportunities that their service and branch control. Such concerns moderate any desire to challenge the conventional wisdom.

The strongest base of support for transformation thus appears to come from the senior ranks. This should come as good news to advocates in innovation, as these officers are already, or will soon be, in positions to effect change. By contrast, it appears that junior officers do not see transformation as something that is important to them. Advocates of change thus need to find ways to mobilize junior officers in support of change. They need to find ways of explaining the advantages of new approaches to combat in very tangible ways. They also need to put in place a system of incentives and rewards to draw the best and brightest young officers into new career paths.

Notes

1 The first survey consisted of 36 statements while the second consisted of 97. Respondents were asked to agree or disagree with each on a scale of 1 to 7, where 1 indicated strong disagreement, 4 uncertainty, and 7 strong agreement. For analytical purposes, we considered answers of 1, 2, or 3 to indicate disagreement with the statement, and 5, 6, or 7 to indicate agreement. We also adopted two different measures of uncertainty: we considered answers of 4 to reflect genuine uncertainty, while those responses with values of 3, 4, and 5 were considered to be tending toward uncertainty.

2 It is worth noting that the statement was phrased somewhat differently in the 2000 survey: "Future adversaries will be able to use long-range precision strike weapons such as ballistic and cruise missiles to destroy fixed military infrastructure, such as ports, airfields, and logistical sites." While this phrasing may account for some of the shift, it is hard to see how it could account for its magnitude.

References

Bruni, Frank (1999) "Bush Vows Money and Support for Military," *New York Times*, 24 September.

Cohen, Eliot A. (1996) "A Revolution in Warfare," *Foreign Affairs*, 75(2): 37–54.

Department of Defense (2001) *Quadrennial Defense Review Report*, Washington, DC.

Department of Defense (2003) *Transformation Planning Guidance*, Washington, DC.

Goure, Leon and Michael Deane (1984) "The Soviet Strategic View," *Strategic Review*, 12(3): 80–94.

Mahnken, Thomas G. and FitzSimonds, James R. (2003a) "Revolutionary Ambivalence: Understanding Officer Attitudes Toward Transformation," *International Security*, 28(2): 112–148.

Mahnken, Thomas G. and FitzSimonds, James R. (2003b) *The Limits of Transformation: Officer Attitudes Toward the Revolution in Military Affairs*, Newport Paper 17, Newport, RI: Naval War College Press.

Office of the Press Secretary (2001) "President Speaks on War Effort to Citadel Cadets," *White House*, 11 December. Online, available at www.whitehouse.gov/news/releases/2001/12/print/20011211–6.html.

Ricks, Thomas E. (2001) "Rumsfeld on High Wire of Defense Reform," *Washington Post*, 20 May, p. 1.

Rosen, Stephen Peter (1991) *Winning the Next War: Innovation and the Modern Military*, Ithaca: Cornell University Press.

Sanger, David E. (2001) "Bush Details Plan to Focus Military on New Weaponry," *New York Times*, 14 February, p. 1.

Weinberger, Sharon (2001) "Joint Force Critical for Transformation Says Head of Rumsfeld Study," *Aerospace Daily*, 29 August, p. 1.

11 Transforming organizational culture

Lessons learned from a systems perspective

Nancy Roberts

No matter what their interest or background, most executives and policy makers consider culture central to the transformation process. Those closely associated with the United States (US) Department of Defence's (DoD) transformation efforts certainly agree. Transformation is fundamentally about changing culture, said Art Cebrowski, former director of the Office of Force Transformation in the Pentagon (Office of Force Transformation 2003). Admiral Giambastini concurs stating, "change in culture is key to joint transformation" (Office of Force Transformation 2004). This view echoes those of Representative McThornberry, who noted that "cultural innovation is needed for successful DoD transformation" (Office of Force Transformation 2004).

Despite the almost universal recognition that cultural change is an important component in the transformation process in military organizations, there has been less agreement on what culture actually is. Most would agree that culture is the "glue" that holds organizations together. Beyond that, some definitions emphasize the more normative aspects of culture – the values, attitudes, and beliefs that leaders espouse; other definitions focus on the actual pattern of behaviour that emerges when organizational members interact with one another, revealing how they formally and informally get work accomplished (Ashkasy *et al.* 2000; Cooper *et al.* 2000; Martin 1992; Martin 2002; Schein 1992).

This chaper introduces a systems perspective that integrates both views. The intent is to illustrate the advantages of a systems perspective, not only for its more inclusive treatment of culture, but also for its utility in helping leaders locate the points of leverage for changing an organizational culture as part of the overall transformation effort. To accomplish this end, this chapter is divided four sections. The first section briefly describes the systems perspective and introduces the organizational systems framework (OSF) model that defines culture as the values, beliefs, and ideals espoused by organizational leadership as well as the pattern of behaviour that emerges as a consequence of the organization's direction and design. The model also identifies where leaders can intervene in a system if they want to pursue cultural change.

Section two enlivens the systems perspective with a case – General Electric's (GE) two-decade transformation. Using the OSF model as a backdrop, the case illustrates how CEO Jack Welch intervened to correct GE's deficiencies, and how, over time, his interventions ultimately produced a dramatic transformation, both in the culture and in the organization's performance. Section three draws out the basic steps in cultural and organizational transformation from the GE case. These generic steps provide a useful outline for any organization embarking on transformation. Drawing on lessons learned from both private and public organizations, section four offers guidance to organizations beginning a transformation process. Focusing on *how* to transform, rather than *what* to transform, it points out the advantages of a process approach to change.

Before beginning the chapter, a caveat is in order. This chapter on cultural transformation focuses on policy implementation – how leaders can intervene in a system to affect a cultural change and what guidance we can give them when undertaking such a task. It does not address the issue of whether cultural transformation should be pursued and to what end. Such questions are matters of policy, and although critically important, are beyond the scope of this paper. Whether leaders should launch cultural transformation depends on a whole set of factors and conditions, not the least of which is the organization's environment, its position within that environment, and the strategy it formulates.

Systems perspective on organizational culture

A system is composed of interrelated elements that form a whole separate from its environment. For example, the human body is composed of elements and subsystems that are interrelated with one another (e.g., the circulatory subsystem, the digestive subsystem etc.). These subsystems, and the elements that constitute them, function as a whole that is greater than the sum of its parts. When combined and interacting, they identify the body as a physiological entity distinct from its environment.

Figure 11.1 illustrates a rough outline of an organizational system. It is composed of direction-setting elements, design elements, culture, and results. Each one of these broad categories can be broken down further into additional elements as shown in the more detailed version of the OSF in Figure 11.2. The organization, its subsystems and their elements, are all embedded in an environment beyond the organization's boundaries. Although the relationship between an organization and its environment is critically important for setting organizational direction and fashioning the design elements, for our purposes in this exercise, the focus is on the interplay *among the internal organizational elements* rather than on the interaction between those elements and the external environment.

The arrows connecting the various elements in Figures 11.1 and 11.2 underscore the interactive aspects of the organizational system. A change in one element impacts the whole. For example, if a leader changes the organization's structure, ripple effects are expected in the other design elements as well as the organization's culture and results. A change in one system element has consequences for all others.

The arrows in the OSF also show feedback loops to illustrate system dynam-

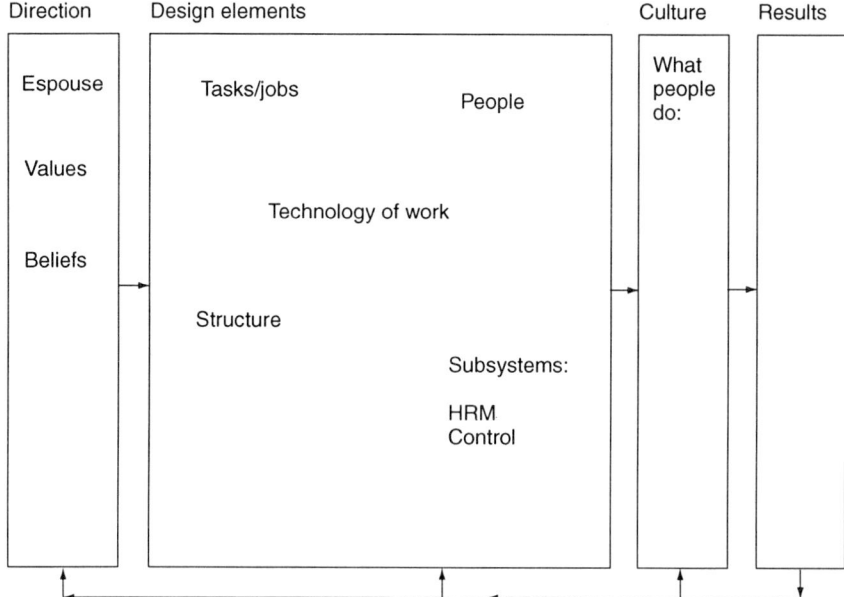

Figure 11.1 Organization system framework.

ics. On the right hand side, the system's results feed back to the organization's direction, design elements and culture. Practically speaking, this means that if the organization's results were poor at time one, we would expect executives to intervene and make changes in order to improve organizational performance at time two. Options for executive interventions and changes are in the organization's direction and/or its design factors. Thus, the OSF has a dual purpose. Leaders can utilize it as a tool to describe and diagnose existing organizational conditions, and they can use it as a mechanism to pinpoint where, and how, to intervene to make improvements in the organization.

A basic assumption of the OSF is that the organization seeks to maintain congruence, fit or alignment, between itself and its environment and among the all the organization's direction and design elements. An optimally functioning organization will achieve a "good fit" among all parts of the system. A poorly functioning organization will be the result of a "poor fit." Thus, the effectiveness and performance of an organization is a reflection of the extent to which there is congruence, alignment, or fit among the key elements (Nadler and Tushman 1998).

Specific to our purpose in understanding the transformation of organizational culture, the OSF incorporates both normative and behavioural views. Espoused values and beliefs (the normative) are part of the direction setting effort that is located on the left hand side of the framework. Along with the mission, vision, goals and strategies, they represent the ideal – what organizational members

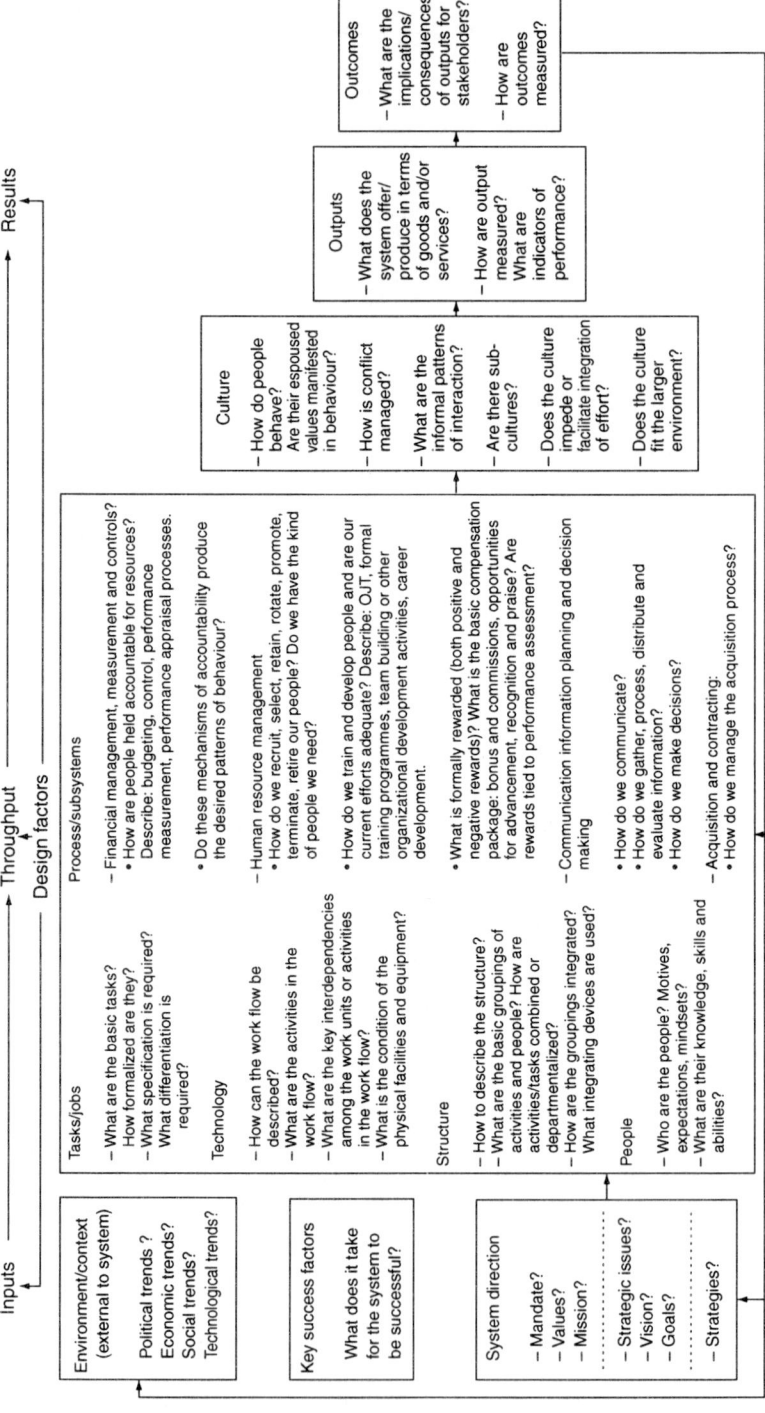

Figure 11.2 Organizational systems framework.

hope to achieve in the future, including expectations of how they plan to interact with one another and their external stakeholders. The second is labelled "culture." It describes what occurs when organizational members actually do interact with others (behavioural). Hence, espoused values, beliefs and ideals are separated from action.

There are advantages in considering organizational culture from this more comprehensive view. First and foremost, distinguishing between what is espoused and what is practised enables us to discover how far apart our ideals are from actual reality. It gets to the heart of the implementation question in organizations – whether what leaders want to happen does indeed occur. Second, we understand from the OSF that culture is considered to be an emergent property of the organization – it evolves as a consequence of the organization's direction and design elements. Its position in the framework underscores the view that culture is not something that can be accessed or controlled directly. To change organizational culture, leaders have to intervene and change other system elements.

The OSF also comes with a warning: cultural transformation is complicated. Organizations are complex, dynamic systems. It is difficult to anticipate first, second, and third-order effects of change, not to mention how organizational elements will interact with the organization's external environment. Although organizational culture is best understood as an emergent property of the system, we do not, as yet, have a clear understanding of the dynamics of this emergence. Interventions can and do go awry; interactions among hundreds of organizational elements can and do produce unintended consequences. Although useful, the systems perspective on organizational culture should be used with some degree of caution. It offers no guarantee that in complex, adaptive systems, interventions to change organizational culture will in fact produce the culture that we desire.

System perspective on GE's two-decade transformation

We now turn to an actual case of large-scale system transformation – General Electric's two-decade change under the leadership of Jack Welch (Bartlett and Wozny 2004; Bower and Dial 1994). Using the OSF as the backdrop to frame the interventions, I summarize how Welch succeeded in transforming GE, including its culture, to achieve results that few organizations have been able to match.

When Welch became CEO in 1981, he set in motion a *new direction* and developed *new ways to measure the direction* – each business unit was to become the number one or number two competitor in its industry or it would be sold or closed down. Between 1981 and 1990, he sold off more than 200 businesses, freeing up over $11 billion of capital, and made over 370 acquisitions. Working toward a more "lean and agile" organization, Welch intervened and made changes in the organization's design. First he instituted a highly disciplined de-staffing process (human resources management) aimed at the headquarters group, including a 50 per cent reduction in the 200-person strategic planning staff. He scrapped the strategic planning subsystem and replaced it with "real time planning" built around a five-page strategy playbook that provided simple one-page answers to five questions concerning the market, the competition, GE's responses to it, the greatest competitive threat over the next three years, and GE's planned response. He changed the organization's structure by reducing the number of hierarchical levels from nine to as few as four, with all business units reporting directly to him. His downsizing, de-staffing and de-layering eliminated jobs, even when factoring in the acquisitions from 404,000 in 1980 to 292,000 by 1989. He also made significant changes in *people*. He created a new "varsity team" of managers with a strong commitment to his values of entrepreneurial enterprise and a willingness to take charge and bring about change. Through these efforts, he began to break the old GE culture, but he still had some distance to travel.

His second stage of transformation began with what he called "software" initiatives – Work-Out and Best Practices. These two, new initiatives focused on *tasks/jobs and the technology of work*. As a forum designed to get unnecessary bureaucratic routines out of the system, Work-Out provided a protected space where employees and their bosses could work out new ways of dealing with each other. Patterned after New England town meetings, groups of 40 to 100 employees met to share views about their businesses and how they could improve them. The rules of the process required managers to make instant decisions on each employee proposal – in front of everyone. By mid-1992, over 200,000 GE employees, two-thirds of the workforce, had participated in the programme. Best practices, launched around the same time, sought to learn from other companies that were achieving higher productivity growth than GE. Welch became such an instant convert to the approach that he committed to a new training programme throughout the organization, and integrated it into the ongoing agenda of Work-Out teams. The focus was then on effective *processes* rather than controlling individual activities, with customer satisfaction (*results*), being the main gauge of performance.

Welch also used *leadership development* as a lever for change. His goal was to realign the mindset and skill set of the company's employees with the new strategic direction. Central to this effort was adapting GE's well-established human resource management (HRM) system. He personally kept close tabs on the company's top 500 executives, using his reviews with them as a time for *coaching and development*. He estimated that he spent at least 70 per cent of his time on people issues. Welch also overhauled GE's compensation package, using stock options as the primary component of management compensation. He also began making more aggressive bonus awards and option allocations tied to individual performance to reward people for change.

Central to his new HRM policies was the management development facility at Crotonville. It became the place where the next generation of leaders would be aligned with his new visions and ideals. The focus was on real priority issues and results-oriented action. Visiting twice a month to teach and interact with employees, Welch used training opportunities to lead the charge for change. He also instituted a rating system to eliminate managers who were not performing – especially those who were not living GE values. "People are removed for having the wrong values ... we don't even talk about the numbers" (Bartlett and Wozny 2004: 8). To reinforce his commitment to the new leadership criteria, he introduced a *360-degree feedback process*. Tied to the *evaluation process*, it became the means for identifying training needs, coaching opportunities, and career planning. He also implemented a *performance appraisal system* that required every manager to rank his or her employees into one of five categories based on long-term performance. This so-called "vitality curve" weeded out poor performers in category five and rewarded those who shared and sought new ideas. He wanted individuals who possessed the 4Es: energy (excited by ideas), ability to energize others, edge (the ability to make tough calls) and execution (ability to turn vision into results).

Welch's third wave of transformation began in the 1990s. At this point he began to focus on "integrated diversity." The company's vision was one of a "boundary-less" organization characterized by an "open, anti-parochial environment, friendly toward the seeking and sharing of ideas, regardless of their origins" (Bartlett and Wozny 2004: 9), reinforcing the earlier initiatives of Work-Out and Best Practices in transforming the *jobs and the technology of work*. It entailed introducing a quick response programme that enabled managers to adopt innovations from outside businesses and to share them across the GE businesses. The company's integration model spread ideas and expertise throughout GE. As Welch wrote:

> The boundaryless company we envision will remove barriers among engineering, manufacturing, marketing, sales, and customer service; it will recognise no distinctions between domestic and foreign operations – we'll be as comfortable doing business in Budapest and Seoul as we are in Louisville and Schenectady. A boundaryless organization will ignore or erase group labels such as "management," "salaried" or "hourly," which get in the way of people working together.
>
> (Bartlett and Wozny 2004: 9)

Welch also introduced the notion of "stretch" in the early 1990s to challenge existing norms and to help set *performance targets*. His objective was to change the way *goals and targets* were set and to *reward* those who achieved them. Central to this effort was the "don't punish failure" concept to enable people to think of fundamentally better ways to do their work without penalizing them if goals were not reached. These efforts and the *new strategic initiative* to move towards service businesses and away from traditional industrial products changed the very nature of the company.

Closing out the decade was Welch's Six Sigma Quality Initiative, a programme designed to reduce the error rate and improve quality in products and services. In order to reach Six Sigma quality levels company-wide by 2000, he began to redesign GE's operating system to ensure greater integration of effort across functional departments. In a carefully orchestrated series of meetings, he announced redesign of GE's *planning, resource allocation, review, and communication* processes, or what is referred to as the organizational processes in the OSF. Observing these changes, Welch was pleased to announce that the programme had begun to "change the DNA of GE to one whose central strand is quality" (Bartlett and Wozny 2004: 13).

Besides improving productivity and organizational performance, the aim of all of these initiatives and interventions to *create a culture* characterized by openness, candour, speed, simplicity, self-confidence, and ability to face reality. Welch admitted these ideals would "take a decade before GE's new culture (would become) as hard to change as the one it is supplanting" (Bowyer and Dial 1994: 13). Yet culture was central in his transformation plan. As he reminded people: "a company can boost productivity by restructuring, removing bureaucracy and downsizing, but it cannot sustain high productivity without cultural change" (Bowyer and Dial 1994: 15).

Welch's efforts over two decades paid off in terms of organizational results. GE's price–earnings ratio climbed from a multiple of seven in 1982 to a multiple of 16 in 1993 – a clear signal that the stock market approved the changes he was making. Over a period when the S & P 500 had risen 36 per cent, GE's market value had grown 498 per cent (Bowyer and Dial 1994: 15). By 1999, GE's revenues exceeded $100 billion for the first time; operating revenues were at an all-time high of 16.7 per cent; and earnings per share had increased 14 per cent over 1997's record level (Bartlett and Wozny 2004). In recognition of this outstanding performance and the company's transformation over two decades, American executives voted GE the country's "Most Admired Company" for the second year running in *Fortune* polls, and the *Financial Times* named it the "Most Respected Company in the World" (Bartlett and Wozny 2004).

Basic steps in cultural transformation

The GE case enables us to extract steps any organization could take to pursue cultural transformation (Table 11.1). Step one begins with a diagnosis of the existing system and includes a description of the organization's direction, its design factors, culture and the results it is currently achieving. The purpose is to understand the system's strengths and weaknesses in order to build on its capabilities and correct its deficiencies.

Welch appeared to follow a path similar to step one when he first took over as CEO. Keenly aware of his system's weaknesses, he moved to "fix, sell, or close" uncompetitive businesses. Scores of businesses were sold: central air conditioning, housewares, coal mining, even GE's well-known consumer electronics business (Bartlett and Wozny 2004). He played to GE's strengths by categorizing the remaining businesses as core (those that had the priority of reinvesting in productivity and quality), as high technology (those that would stay on the leading edge by investing in R & D) and as services (those required to add outstanding people and make contiguous acquisitions) (Bartlett and Wozny 2004).

Compatible with step two in Table 11.1, Welch set the general direction by insisting all retained GE businesses become either the number one or number two competitor in their industry. Acquisitions then followed to support this mix of businesses. He avoided establishing an overall theme or single strategy due to GE's large portfolio. Instead, he made it clear he wanted a profitable, highly diversified company with world-quality leadership in each of its product lines. Central to this overall effort was to reduce GE's dependence on its traditional industrial products. Overtime, more and more emphasis was on services where he believed the opportunity was unlimited. As he noted, "We have to change the very nature of what we do for a living" (Bartlett and Wozny 2004: 11).

In addition to the new direction he set, Welch espoused new values and ideals consistent with step three in Table 11.1. He wanted GE to be a "unique, high-spirited, entrepreneurial enterprise," more "lean and agile." Competitive people were needed to take charge and bring about change. The ideal was to create the culture of a small company – a place where people were not afraid to have voice but instead could be direct and honest with one another. Open engagement, participation, and cross-boundary cooperation were essential to successful problem solving and decision-making.

At the same time, Welch was quite clear what aspects of the old system and

Table 11.1 Basic steps in cultural transformation

1   Describe existing organizational system in terms of its strengths and weaknesses. Specify where (in terms of its existing direction, design factors, culture, and results) its capabilities and deficiencies lie.
2   Set new strategic direction and establish what results you want to achieve to take advantage of your capabilities and correct your deficiencies.
3   State your ideals and values. Describe the culture you want to create.
4   Locate points of intervention in the organization where change needs to occur in order to support your new direction, espoused values and beliefs. Consider the following points of interventions in the design elements:
   a   Change of Structure.
   b   Change of Jobs/Tasks.
   c   Change of Technology of work.
   d   Change of People.
   e   Change of Organizational processes and subsystems:
       HRM
       Controls, Measurement
       Planning, Communication, Decision Making
5   Design interventions into the design elements that are consonant with the your ideals and the new culture you hope to create.
6   Monitor your performance over time:
       Did you achieve the results you wanted?
       Were your interventions to change the organization's design successful?
           Are your espoused beliefs and values consistent with how people behave?
           Has a new culture emerged to replace the old culture?

culture he would not support. He rated top managers on the extent to which they "lived GE values" insisting that:

> There was no place at GE for the adherents of the old culture: We take people who aren't boundaryless out of jobs.... If you're turf-oriented, self-centred, don't share with people and aren't searching for ideas, you don't belong here.
>
> (Bartlett and Wozny 2004: 9)

Welch's interventions to change the organization's design were comprehensive. Over a 20-year period, although he chose to focus on different design elements at different points in time, he included all elements in his transformation package (See Table 11.1, step four). Welch restructured the organization; redesigned tasks and jobs and the technology of work; recruited new people and developed the skills and competencies of those who remained, putting a heavy emphasis on training and development. He completely revamped all the organization's subsystems and processes including HRM (hiring, selection, evaluation, development, promotion, and rewarding of personnel), controls and measurement, planning, communication and decision-making. He left nothing to chance.

Consistent with step five, Welch also sought to make his change interventions compatible with the GE culture he wanted to create. With the exception of the early years when he earned the nickname "Neutron Jack" for his drastic restructuring, he designed initiatives to support both the increase of productivity and cultural change. For example, the purpose of Work-Out was two-fold – get unnecessary work out of the system and at the same time provide a forum in which people (bosses and employees) could create honest, energetic interaction, build trust, and practise new ways of dealing with each. As Welch elaborated:

> Work Out has a practical and an intellectual goal. The practical objective is to get rid of thousands of bad habits accumulated since the creation of General Electric.... The intellectual part (redefines) the relationship between boss and subordinate. I want to get to a point where people challenge their bosses every day.... Now, how do you get people communicating with each other with that much candour? You put them together in a room and make them thrash it out.
>
> (Bower and Dial 1994: 12–13)

The Best Practices Initiative had a similar purpose. As another assault on business-as-usual, it sought to increase productivity by comparing GE with other high performing organizations that had sustained their productivity growth for at least ten years. GE learned that almost every company emphasized managing processes, not functions; focused less on individual department performance and more on how they work together as products move from one to the other. From its studies, GE realized it was measuring the wrong things. Rather than focus on *what* should be done, the emphasis needed to shift to *how* things could be done, how people could work together across boundaries to improve organizational effectiveness. The initiative was then embedded in the Work-Out sessions to make cross-boundary collaboration a reality.

Finally, in line with step six in Table 11.1, GE was careful to monitor its performance. Welch could always point to results, both in terms of GE's overall performance and in terms of each initiative it pursued. For example, the Six Sigma Quality Initiative, introduced in the 1990s to reduce error rates, far exceeded anyone's expectations. In the first two years of operation, although $500 million had been invested to train 5,000 managers and the entire professional workforce of 85,000, returns of $750 million over the investment were achieved. The forecast for additional returns was $1.5 billion for 1999 alone.

Measuring cultural shift is more difficult than measuring productivity and organizational performance, but even in this instance, Welch had a response to an interviewer who asked: "When will we know (your) changes have worked?" He answered:

> Ten years from now, we want magazines to write about GE as a place where people have the freedom to be creative, a place that brings out the best in everybody. An open, fair place where people have a sense that what they do matters, and where that sense of accomplishment is rewarded in the pocketbook and the soul. That will be our report card.
>
> (Bartlett and Wozny 2004: 7)

Process guidance on cultural transformation

What lessons can we derive from the GE case that would inform organizations embarking on transformation, especially those concerned with transforming their cultures? Having outlined *what* has to change in previous sections, our attention now turns to the process of change – *how* leaders should approach the transformation effort.

Guidance begins with the acknowledgement that organizational size is a factor. The larger the organization, the greater is the challenge. In large, complex bureaus, instead of attempting transformation in all system elements at the same time, one is advised to follow the path that Welch established in GE. Take an orderly approach to the change process and introduce transformation in phases. Decide what needs to be changed first, and what system elements should follow, and in what sequence. Be clear with organizational members and stakeholders what is to be accomplished in each phase. Viewing transformation in stages avoids the trap of promising and expecting too much in a short period of time.

This approach demands a very sophisticated level of change agency. It puts a heavy emphasis on the leader's ability to diagnose a system's weaknesses, set a new direction, and design change interventions for each phase of the transformation process, anticipating what the organization will need at each point in time. Most importantly, the leader has to be sensitive to how far to push the organization at each stage. Attempting too much, too quickly can overwhelm a system. Doing too little can derail momentum. Knowing how much and how far to push an organization is an art, although Welch made it clear where he stood on the matter:

> Changing the culture starts with an attitude. And I would suggest it starts at the top with the CEOs and the boards of directors that are charged with leading our institutions. More boards have to be thinking: how much can this organization take, how much can it absorb, is it being stressed too little or too much – constantly challenging the pace. How does an institution know when the pace is about right? I hope you won't think I am being melodramatic if I say that the institution ought to stretch itself, ought to reach, to the point where it almost comes unglued.
>
> (Bower and Dial 1994: 6)

In designing a transformation process, one also is advised to factor in how radical the change will be. Making marginal changes (incremental change) in an organization's planning process is vastly different from changing the total system. Radical change reconfigures all the system elements to form a new whole. It requires interventions in both direction setting and all the design factors. To adequately prepare for a change process, and to summon resources and people who will be needed for this undertaking, it is important to know whether one is dealing with radical or incremental change. Welch understood the distinction between the two, although for him the idea was: "shun the incremental and go for the leap" (Bower and Dial 1994: 5).

In addition, it is wise to attend to people's head, hands, and hearts when embarking on transformation. In the best case, change interventions need to be designed with all three in mind. As Jeanie Duck (1993: 109) counsels, "change is intensely personal. For change to occur in any organization, each individual must think, feel, or do something different." People need to understand conceptually what is expected of them (head); they need to know how they will work differently during and after the change (hands); and they need to be emotionally connected and engaged in support of the change effort (hearts).

Welch recognized and used all three levels of engagement. He was relentless in explaining what needed to be done and why, providing a context and rationale for change. He ensured that people had the resources for training and education so they could learn and practise new ways of problem solving and working together. He constantly communicated and engaged people on a personal level to win their commitment to the change process. He offered this advice on how to make more of a direct connection with people:

> Good business leaders create a vision, articulate the vision, passionately own the vision, and relentlessly drive it to completion. Above all else, though, good leaders are open. They go up, down, and around their organizations to reach people. They don't stick to the established channels.

They're informal. They're straight with people. They make a religion out of being accessible. They never get bored telling their story.

Real communication takes countless hours of eyeball to eyeball, back and forth. It means more listening than talking. It's not pronouncements on a videotape; it's not announcements in a newspaper. It is human beings coming to see and accept things through a constant interactive process aimed at consensus. And it must be absolutely relentless.

(Bower and Dial 1994: 6)

Efforts to take a more holistic approach to transformation, engaging head, hands, and heart, are evident in one technique the US Navy is using to transform its culture. The Navy has introduced Appreciative Inquiry (AI), a new large group intervention technique that can bring hundreds of people together at any one time. The essence of Appreciative Inquiry (AI) is a collaborative search to identify and understand the organization's strengths, its potential, its greatest opportunities, and people's highest hopes for the future. The intervention begins when participants from all levels within the organization collect and share stories about the organization's "positive core." These personal accounts document not only what has happened in a particular situation, but they also explore how people *felt* about the event as they experienced it. Participants then ask themselves how they can build on these positive experiences and make them more of the norm throughout the organization. They then collectively identify and launch projects in the organization more in keeping with these positive ideals. Reports from these AI events point to a rising level of energy, commitment and enthusiasm for those who have experienced them as part of the Navy's change effort.

We must remember too that change interventions of this type are "living laboratories" and can be used for multiple purposes. Welch used the change initiatives of Work-Out and Best Practices to introduce new cultural values and afford people an opportunity to practice them. Each initiative was an occasion to espouse the values of candour, honesty and energetic interaction. And each initiative gave participants the occasion to demonstrate how they could act with candour, honesty and energy. The Navy's leadership summits have similar goals. Leadership both espouses the importance of "breaking down its stovepipes" and it gives participants time to practise it during its AI meetings and in the cross-boundary collaboration required as part of follow-on activities.

Keen observers of organizational change programmes encourage this practical orientation toward change, especially those that focus on people's emotions. As Jeanie Duck (1993: 16) notes, "change is fundamentally about feelings." "Companies that want workers to contribute with their heads and hearts have to accept that emotions are essential to the new management system. … Managing people is managing feelings" (Duck 1993: 113). Unfortunately, as numerous studies reveal, some of the feelings about organizational change are not positive. Organizations are "full of 'change survivors', cynical people who have learned to live though change programs without really changing at all"

(Duck 1993: 112). Instead of commitment, they have complaints. Instead of belief, they have doubts; "I'll believe it when I see it" (Duck 1993: 112), they retort. For them, efforts to change are "just another management fad in an endless series of management fads" (Duck 1993: 112).

Highlighting the positives and keeping people energized, focused, and rewarded for change are multiple ways to prevent transformation fatigue and the self-defeating cycle of fear, cynicism, and resistance from derailing initiatives. Most studies on organizational transformation recommend these and other steps to increase the probability of success (Beer and Nohria 2000; Beer *et al.* 1990; Duck 1993; Kotter 2000; Pascale *et al.* 1997). Unfortunately, despite these broadly established principles found in the literature, we have evidence that 50 per cent to 70 per cent of all change efforts fail (Beer and Nohria 2000; Kotter 2000). Worse still, we know that organizational performance in the short term is often disrupted and degraded (Beer *et al.* 1990; Duck 1993; Pascale *et al.* 1997).

Here is where leadership comes into play. Leadership is central to cultural transformation according to Edgar Schein. "Organizational cultures are created in part by leaders, and one of the most decisive functions of leadership is the creation, the management, and sometimes even the destruction of culture" (Schein 1992: 35). Unfortunately, according to John Kotter (2000) of the Harvard Business School, leaders are not often up to the task because they do not establish a great enough sense of urgency; do not create a powerful enough guiding coalition; lack a vision; under communicate by a factor of ten; do not remove obstacles, both human and systemic to their vision; do not plan for and create short term wins; declare victory too soon; and do not anchor changes in the corporate culture.

From Schein's perspective, the key to successful leadership is telling the truth. Leaders face unpleasant facts whether it concerns the competition, decreasing market share, lack of revenue, or stakeholder discontent. As Welch pointed out, leaders need to face reality, be candid, and see "the world as it is rather than as you wish it were" (Bower and Dial 1994). As he pointed out, "candid managers – leaders – don't get paralysed about the 'fragility' of the organization. They tell people the truth. That doesn't scare them because they realize their people know the truth anyway" (Bower and Dial 1994: 6).

Truth telling in response to organizational realities requires leaders to develop a picture of the future that clarifies the direction toward which the organization needs to move. Central to these efforts is communication. Leaders must listen to others and they must use all existing communication channels to communicate with others in both words and deeds. As John Kotter notes:

> Transformation is impossible unless hundreds or thousands of people are willing to help, often to the point of making short-term sacrifices. Employees will not make sacrifices, even if they are unhappy with the status quo, unless they believe that useful change is possible. Without credible communication, and a lot of it, the hearts and minds of the troops are never captured.
>
> (Kotter 2000: 63)

Final process guidance on cultural transformation centres on getting expert support for leaders launching a transformation effort. Unless they have had the good fortune to participate in successful programmes and watch at close hand how individuals like Jack Welch operate, most leaders and managers have witnessed more failure than success. What they need is a framework, a mental map to guide them and their organizations (Harrison and Stokes 1992: Schein 1999). Practically speaking, this may mean getting assistance from experts who can provide the frameworks and coach them through the transformation process. Experts knowledgeable in the use of new social technologies are especially helpful.

New social technologies, such as Appreciative Inquiry, Future Search, and The Search Conference, are designed to accommodate large numbers of people and bring the "whole system in the room" (Bunker and Alban 1997; Emery and Purser 1996; Weisbord and Janoff 1995). They enable organizations to create new visions and organizational designs more expeditiously and with greater organizational participation than other change techniques. Although their application requires a high level of skill and expertise, they are very useful tools in the transformation process and offer great promise for those embarking on change. One recent example, "Listening to the City," www.americaspeaks.org, attests to their power.

"Listening to the City" involved nearly 4,500 people in a day-long session to discuss and recommend which plan should be chosen to reconstruct the World Trade Centre in New York City. It employed experts, the America Speaks Organization, to design and manage the conference, invite participants, facilitate discussion, and collect and analyse participants' responses to a series of questions. The organizers utilized the latest advances in information technology to assist them in their tasks: they gave every participant a keypad to register his or her views; they used computers to process data in real time so people knew, within minutes, the results of the facilitated discussions in hundreds of ten-person groups. By the end of the day, 4,500 people presented a unified proposal and recommendations on the reconstruction of the World Trade Centre to New York officials. Watching the events unfold, even the most sceptical observers were convinced that a group-based approach to change that included thousands of people could be successful, as long as it was expertly designed, supported, and facilitated.

Conclusion

Transformational change is complex, difficult, and fraught with risk. The goal here has been to minimize the complexity, difficulty and risk by introducing the Organizational Systems Framework to guide leaders through the change process. Using GE's very successful case of organizational transformation as a backdrop, it has offered a map to help pinpoint *what* changes need to be made, what *steps* or phases could be followed, and *how* leaders could operate using *process guidelines* to avoid the worst "landmines" in transformation work.

This general overview has not addressed specific applications to military organizations, although militaries around the world are keenly interested in the topic and many of them have transformation programmes underway. Unfortunately, it is too early to assess the impact of these worldwide initiatives. What is possible at this juncture is to offer a few observations from a systems perspective on one organization undergoing transformation, one with which I am most familiar – the US Department of Defense under the leadership of the Secretary of Defense, Donald Rumsfeld.

Initial transformation in DoD has focused on the Revolution in Military Affairs. Translated into OSF terms, the thrust of this revolution has been on changing the *technology of work* and the *tasks or jobs* that are required to conduct field operations. Evidence of this revolution has surfaced when, for example, men on horseback call in air strikes in Afghanistan, and new information technology used in Iraq gives foot soldiers a more complete picture of the battlespace. DoD leadership has made a strong case to intervene into these organizational design elements given the new US threat environment. From a systems perspective, military operations are a logical place to drive transformational change.

Yet transformation requires realigning all system parts to form a new whole. To what extent are the other system elements being realigned to support this revolution in the technology of work? Is there a comparable "Revolution in Business Affairs" to reconfigure the other design elements to be congruent and supportive of the changes in military operations?

Unfortunately, the answer to this question, at least at this point in time, is no. Although, there have been efforts to reform key design elements, such as acquisition and contracting subsystems, past and current initiatives have not been well integrated to match changes in the new technology of work on the operational side. It does no good to design operations around new technology if the acquisition and contracting processes are so slow and cumbersome that they cannot deliver the technology in a timely manner to those who need it. The same argument holds for financial subsystems that delay resource allocation, human resource subsystems that do not recruit, train, reward, and promote personnel in line with transformation goals and the new technology of work, organizational structures that impede information flow and encumber decision making with layers of hierarchy, and controls that focus on inputs rather than on results.

One could argue that changing a design element, such as the technology of work, will eventually produce ripple effects throughout the entire system, so ultimately, adaptations in other system elements will gradually occur. There are dangers in this more incremental approach to change, however. The pace of change would be slower, and potentially in the interim, create serious system misalignments and incongruities. The congruence model warns us to expect degraded performance when system elements do not fit together well. There is the additional danger that bureaucracies, taking advantage of a slower pace, will continue their routines and standard operating procedures, unwilling to make necessary adaptations. "Resistance to change" is a well-know phenomenon in the change literature. In whatever guise it takes, it can and does derail change projects, and there is no reason to believe that it would not be a factor in DoD organizations.

From a systems perspective, unless and until there is a more coherent, coordinated and systematic drive to realign all organizational design elements and subsystems to support the Revolution in Military Affairs, it will be difficult to deliver on Rumsfeld's vision of a new military for the twenty-first century. Working on a maximum eight-year time horizon, in a system far more complex system than GE, Rumsfeld faced enormous challenges in building a foundation of sufficient scope and depth to sustain the momentum for transformation after he leaves. But even if he was successful during his tenure, it remains to be seen whether future secretaries of defence will share and support his vision for the US military. As we learn from the GE case, without relentless, energetic and visionary leadership, organizational transformation is not likely to succeed.

References

Ashkasy, N., Wilderrom, C., and Peterson, M. (eds) (2000) *Handbook of Organizational Culture and Climate*, Thousand Oaks, CA: Sage.

Bartlett, C.A. and Wozny, M. (2004) *GE's Two-Dimensional Transformation: Jack Welsh's Leadership*, Boston, MA: Harvard Business School.

Beer, M. and Nohria, N. (2000) "Cracking the Code of Change," *Harvard Business Review*, May/June 2000, 78(3): 133–141.

Beer, M., Eisentat, R.A. and Spector, B. (1990) "Why Change Programs Don't Produce Change," *Harvard Business Review*, Nov/Dec 1990, 68(6): 158–166.

Bower, J.L. and Dial, J. (1994) *Jack Welch: General Electric's Revolutionary*, Boston, MA: Harvard Business School.

Bunker, R.R. and Alban, B.T. (1997) *Large Group Interventions*, San Francisco: Jossey-Bass.

Cooper, C.L., Cartwright, S. and Earley, P.C. (eds) (2000) *Organizational Culture and Climate*, New York: John Wiley & Sons.

Duck, Jeanie D. (1993) "Managing Change: The Art of Balancing," *Harvard Business Review*, Nov/Dec1993, 71(6): 109–118.

Emery, M. and Purser, R. (1996) *The Search Conference*, San Francisco: Jossey-Bass.

Harrison, R. and Stokes, J. (1992) *Diagnosing Organizational Culture*, San Francisco: Jossey-Bass.

Kotter, John (2000) "Leading Change: Why Transformation Efforts Fail," *Harvard Business Review*, March/April 1995, 73(2): 59–67.

Martin, Joanne (1992) *Cultures in Organizations: Three Perspectives*, Oxford: Oxford University Press.

Martin, Joanne (2002) *Organizational Culture: Mapping the Terrain*, Thousand Oaks, CA: Sage.

Nadler, D. and Tushman, M. (1998) *Strategic Organization Design: Concepts, Tools and Processes*, Glenview, IL: Scott, Foresman and Company.

Office of Force Transformation (2003) "Inside the Pentagon: Cebrowski Calls for Cultural Changes," *Office of Force Transformation, Department of Defense*, April, online available at www.oft.osd.mil/library/library_files/article_42_Inside the Pentagon-story from Flash conference.doc.

Office of Force Transformation (2004) "Giambastiani: Change in Culture Key to Joint Transformation," *Sea Power*, September, online available at www.oft.osd.mil/library/library_files/article_410_Sea Power.doc.

Pascale, R., Milleman, M., and Gioja, L. (1997) "Changing the Way we Change," *Harvard Business Review*, Nov/Dec 1997, 75(6): 126–139.

Schein, Edgar (1992) *Organisational Culture and Leadership*, San Francisco: Jossey-Bass.

Schein, Edgar (1999) *The Corporate Culture Survival Guide*, San Francisco: Jossey-Bass.

Weisbord, M. and Janoff, S. (1995) *Future Search*, San Francisco: Jossey-Bass.

Index